SELECT HOUSE

精品屋

赵婷婷 编

CONTENTS
目录

CONTENTS
目录
002

PREFACE
前言
004

"360"INNOVATIVE LIVING
"360"创新生活
016

ALMA ROAD RESIDENCE
阿尔玛住宅
024

APARTMENT S
S公寓
030

ANGLER'S BAY SHOW FLAT
海云轩
038

BEIJING CHATEAU TOWERS
北京公馆
046

BOHEMIAN EMBASSY BEDROOM MODEL SUITE
波西米亚大使馆套房样板间
054

CAMPOS ELISEOS APARTMENT
伊莉斯草原小屋
060

GREEN PENTHOUSE LOFT
绿色小棚屋
066

HARDWICK TUNBULL BEACH HOUSE
哈德维克通布尔海景房
074

HSIEH'S RESIDENT
谢公馆
082

KENSINGTON HOUSE
凯盛顿公寓
092

KENSINGTON TOWNHOUSE
凯盛顿联排别墅
098

LAS LOMAS
洛马斯洋房
104

LOFT IN MADRID
马德里公寓
112

LA ROSSA
映岸红
120

MILLENNIUM TOWER LOFT
千年塔阁楼
129

目录
Contents

MANGROVE WEST COAST 136
红树西岸

METROPOLITAN CHIC 144
美丽都市

NATURAL ORGANIC SPACE 152
自然有机空间

ORCHARD TURN SHOWFLAT 160
乌节弯公寓

PEAK HOUSE 166
现代几何建筑美

PENTHOUSE SCHRADER 172
Schrader豪华公寓

PRIVATE RESIDENCE TIP TOP LOFTS 180
豪华公寓

QUANT# 188
船桨之家

RED 198
红

RESIDENCE TSAI'S 204
郭南国美蔡宅

SOUTH BAY 212
南湾

SPACE OF FEMININITY 218
女性居家环境

SPRING HILL HOUSE 226
春山之屋

STONEHEDGE RESIDENCE 234
斯通家园

TOWN LANE RESIDENCE 242
乡间小巷住宅

UPDOWN COURT SHOWFLAT 252
爱敦阁样板房

WINDSOR LOFT 262
温莎阁楼

INDEX 270
索引

Preface
前言

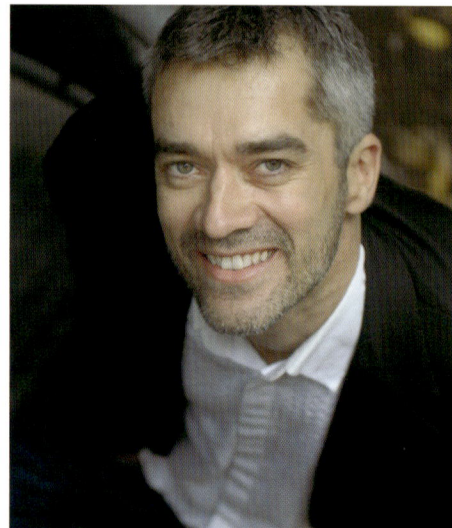

René Dekker
Head of Residential Interiors
SHH

雷内·德克尔
SHH设计公司——主席设计师

Past, Present and Future:
History and Trends in Interior
Design

过去，现在和未来：
室内设计的历史及发展趋势

The sheer diversity of the scope of present-day residential interior design is unparallelled in history. While mainstream publications provide us with snapshots of the most fashionable and cutting-edge elements in interior design, alerting us to what is hot and what is not, there is also a wide variety of more specialised media, tempting us with an even broader spectrum of what is going on around the world – not to mention the proliferation of globalised information available on the internet.

As short as thirty years ago, most interiors would have been "local" – created by local designers for local clients with local materials, factories, artisans and consultants. Only the fortunate few have experienced the luxury of creating a foreign or exotic scheme. Now, almost everything is attainable and styles are as varied and multifaceted as imaginations and budgets could be!

Once upon a time, choice was limited by geography and the small number of creative individuals in the field, and of course by technology. From early records we know that the wealthy and elite of great civilisations such as the Romans and the Greeks had already started to design their interiors. Great care and thought were placed into layouts of homes, but no expense was spared when furnishing them. There can be no doubt that these interiors were rich and colourful – filled with art and sculpture or with intricate mosaic flooring – as they could be in that time. In recent years, technological

当今，家居室内设计出现了前所未有的多样化。市面上的主流书籍为我们展现了最为时尚尖端的设计风格，提醒人们哪些正新潮，哪些已落伍。同时还有大量的专业媒体给我们提供了广阔的视角领略世界各地发生的事件，更勿用说互联网上广泛传播的全球信息了……

在短短的30年前，室内设计还很本土化，由当地的设计师为当地的客户设计，使用的也是当地的材料、工厂、工人和顾问。只有极少数的幸运儿才设计过具有异域风情元素的作品。如今，万事皆备，设计风格可以随着人们丰富的想象力和不同的预算开支而呈现出多样性。

过去，设计创作除了受到地域和技术的限制，还缺乏创造型人才。据记载，在古罗马、古希腊，已有富人和精英开始设计自己的住宅了。他们对房间内的陈设大费脑筋，装修时也绝不吝啬。这些室内陈设富丽堂皇，色彩丰富，有艺术品、雕塑和图案复杂的马赛克地板。不过当时可供选择的材料十分有限。纺织品也局限于天然的材质，如真丝、羊毛、亚麻、棉布等，只做了简单的编织和染色。家具作工虽十分精巧，但其舒适性却差强人意。

advances have created unbelievable options in fabrics and furniture, about which everything is possible, with design and structure limited only by people's imagination. With the Industrial Revolution and the mass production of fabrics and furniture, new designs proliferated, which in turn inspired creative minds to push the boundaries of conventional thought. Suddenly you could have an Egyptian or Moroccan interior, even if you had never travelled there before. Mass production not only created cheaper goods, but also brought in foreign delights. It is now possible to own a clock made in Germany, a rug woven in Belgium and an armchair made in Britain!

Progress and improvement had been exponentially made in the following years, with the latter half of the 20th century seeing a literal explosion of diversity in the design world. The invention of man-made materials – plastic, for example – led to more refined products such as vinyl and polyester. Those were initially used to produce crude copies of leather and silk, but now, in the 21st century, they have blossomed into spectacular elements in their own right. MDF and engineered wood have led to the realisation of previously impossible ideas and the age of computer has offered us unlimited ways of producing and reproducing extremely intricate designs. Adding in the advent of laser, one feels as if anything is now truly possible.

近年来，随着技术的进步，人们在纺织品和家具方面的选择多得惊人，没有实现不了的要求，唯一对设计有所限制的是人类自身的想象力。工业革命后，纺织品和家具都采用大批量生产，新的设计式样不断增加，从而要求设计师的思想突破传统的局限。你能让自己的家变成埃及或摩洛哥风格，虽然你从没去过这些地方。大批量生产不但降低了产品价格，还带来了国外的好东西。人们可以买到德国的时钟，比利时的地毯和英国的扶手椅！

在接下来的几年里，设计水平有了飞速发展和提高，到了20世纪下半叶，设计元素变得更加多元化，随着塑料的发明，继而出现了乙烯和聚酯等人造材料。这些材料最初仅用于仿造皮革和真丝，到了21世纪的今天，已衍生出大量新型材料。中密度纤维板和工程木材将以前不可能的构想变成现实。电脑创造了无限的生产力，让我们能进行更为复杂的设计。再加上激光的出现，一切设想都能实现。

In the late 1980s and early 1990s, the stage was set for the bright and vibrant colours of Tricia Guild's "Designers Guild", heralding a fresh and new concept for the world of design and overturning the basic colour palette of the previous decade. After years of dominance, this was followed by the calm and tranquil palette of designer Kelly Hoppen as the "Mode du Jour", which in turn was followed by the recent penchant for everything crystal. Even though these styles were diametrically different, they also share some elements in common, such as pattern, which has always been drawn from history and which helped to thread the styles together in an invisible patchwork.

Just as the case in the recent past is, historical features have always played an important role in shaping current trends and it is no more so than now. Baroque and Rococco elements have been stylised and altered to suit modern taste with their wonderful shapely curves featuring in everything from wall coverings to furniture, but with a difference. The wall covering is no longer paper and is now embossed instead of printed and is translucent rather than opaque!

时尚潮流都是周而复始的，但每次轮回又都更加精致并趋于完善，可能增加了新的颜色或其他过去没有的元素。20世纪80年代末到90年代初，特里西娅·盖尔德创立的"设计师盖尔德"品牌让鲜艳明快的颜色在设计界风靡一时，这是一个全新的设计理念，颠覆了过去几十年的单一色调。数年后，这种风格又被设计师凯丽·霍朋的"朝夕之间"品牌平静安宁的风格所取代，到了最近又开始趋向使用水晶等发亮的材料。这些风格虽然各不相同，但他们的设计格局是一样的，都是从历史中取出值得借鉴的元素，再将它们拼在一起。

虽然设计的项目已成为过去，但古老的设计风格仍对目前流行趋势的形成起着巨大的作用。巴洛克和洛可可风格已成为固定模式，为适应现代人的审美观，稍稍做了些许改变。从墙壁到家具，美丽的曲线无处不在，却与传统形式略有差别。墙壁上不再使用壁纸，而改为半透明的浮雕式墙绘！

Our world has become smaller. Travel allows many people – not just the rich – to experience exotic places and allows the brave few to wander off to explore more obscure areas, for the discovery of traditional craftsmanship. What started with tradespeople buying great quantities of these crafted products and selling them on the western markets led to the new idea that pre-eminent designers could combine traditional manufacturing methods with Western-European styling to create cutting-edge designs. For example, exquisite glass lighting is being produced in countries such as the Czech Republic, with centuries-old methods, but with a new lust for life. Gone are the fussy details and in their place are chic clean lines. Countries such as India are combining natural elements such as mother-of-pearl with traditional workmanship to create contemporary products with unbelievable finishes that are not only exquisite, but durable and serviceable. And, of course, China, whose ancient designs remain inspirational and who for many years helped produce the goods that fuelled western society, has finally come to the modern era by providing us with affordable top-quality products that emulate the standard of the style and content of previously unaffordable design elements.

我们的世界变得越来越小。旅游不再只是富人的活动，人人都可以体验异国风情，勇敢者还能去探寻未开发的蛮荒之地，发扬传统的手工艺技能。商人开始大量购买这些手工艺品，并销往西方市场，卓越的设计师将传统制造方法与西欧风格相结合，创造出顶尖的设计。比如，精美的水晶灯是捷克的特产，其工艺虽有几百年历史，却是设计师们的新宠。他们摒弃了其中过于繁琐的细节，将其塑造成一道室内的美丽风景。印度等国家自然条件优越，以传统工艺制作的珍珠产品，不仅精美绝伦，而且持久耐用。当然，中国古老的设计风格仍然是众多设计师灵感的来源，多年来中国的产品一直充斥着西方市场，如今终于走向成熟，为我们带来高品质低价格的产品，紧跟时尚潮流，价位又不像大品牌那样高不可攀。

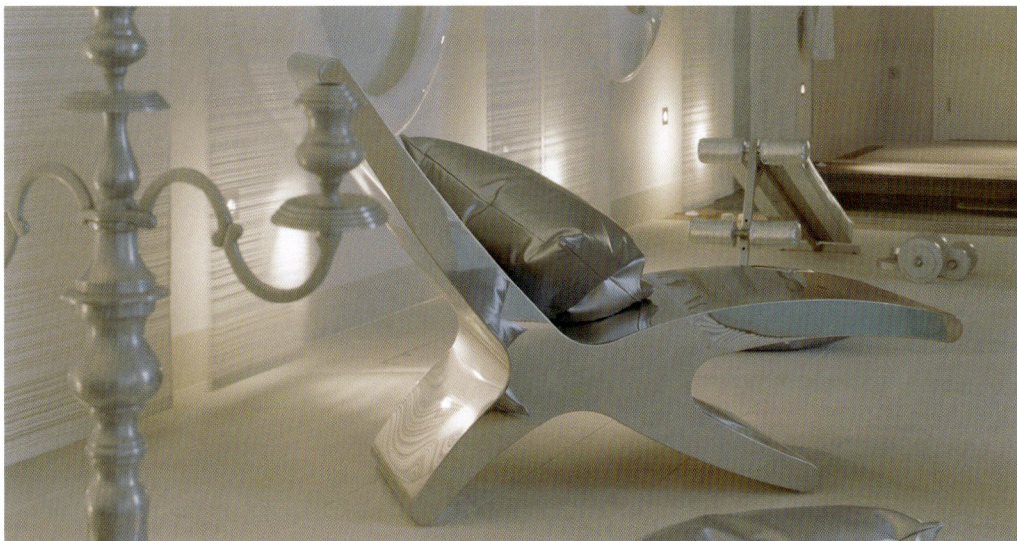

At the opposite end of the spectrum there is also a new generation of designer, not only searching for fresh and exciting products, but creating them. Combining up-to-the-minute technology with traditional design, individuals such as Tord Boontje and Marcel Wanders at Moooi have overturned our concept of what is hip and fashionable. Stunning products such as the "Garland" light and the "Sky Garden", not to mention the ubiquitous "Light Shade Shade", are all testament to the fact that, although you can't reinvent the wheel, you can certainly remodel it! Pushing the boat out even further are inspiring people such as Karim Rashid and Ron Arad, who strive not only to provide the world with brilliant and innovative new products and designs, but also, and especially, to impart their creative knowledge by teaching and inspiring the young talent of tomorrow.

Lighting is especially an area where products have developed in leaps and bounds. The German company, Windfall, reinvented the traditional crystal arm chandelier by constructing the Balance, a free-floating, totally transparent version of its former self. Not only does each candle suspend from the ceiling on its own invisible thread, but the light source is neatly hidden in the ceiling, making this entire visual feast appear ghostly and ethereal. Equally beautiful is the range of candle lights designed by American Kevin Reilly, which emulate the style and romance of a bygone era by creating what appear to be traditional bronze patinated sconces and candelabras housing the humble candle, but actually, the candle is electrically illuminated and made from a modern material which will hold its shape and form for the life of the fitting!

相反的，还有一批新生代的设计师，不仅寻找令人激动的新鲜事物，还自己动手来创造。"莫依"品牌的托德·波提和马可·温德斯将最新技术与传统设计相结合，颠覆了我们对时尚的理解。"花环"灯和"空中花园"等令人惊叹的作品，还有无处不在"光与影"，都证明了一个事实，虽不能另辟蹊径，却能重新诠释。卡里姆·拉什德和罗·阿拉德在这条道路上走得更远。他们不仅致力于创造充满智慧和创意的新产品和新设计，而且尤其是后者，还将自己的创造性思维传授他人，帮助年轻人成才。

灯具能大幅度提升空间品质。"横财"是一家德国公司，他们改造了传统水晶吊灯的平衡方式，营造出全透明吊臂的飘浮效果。每根烛形灯都以隐形的线固定在天花板上，光源也恰到好处地隐藏在天花板之中，看上去飘逸而神秘。美国设计师凯文·赖利设计的蜡烛照明系列有着同样的美感。它仿照浪漫的古代烛灯，蜡烛下面是传统的青铜色烛台。实际上，蜡烛是用电的，以一种现代材料制成，外形足以以假乱真！

In our haste to lionise the breakthroughs of contemporary design, we should not forget to venerate those established designers who have maintained long and successful careers by providing a combination of classic design and quality service. Nicholas Haslam, Alberto Pinto and Jacques Garcia are just a few of the talented individuals who are and always will be the backbone of residential interior design. They also have a niche in the broad spectrum of design, not necessarily making use of the latest trends but concentrating rather on the classic, whilst enjoying the advances in manufacture. They might prefer an 18th century pattern, but would redefine it by way of a 21st century fabric.

Finally, what about bad designs? Is there anything left in this new design-conscious world of ours? The answer of course is yes, but according to whom? Mass production has also brought with it watered-down designs, which cut corners to make products cheaper and more accessible to the world, but are they bad designs if the general population is buying them and who is to blame? In our mission to beautify the world, have we, the designers, created a monster? The proliferation of publications, books and TV shows all dedicated to residential design has made the public more aware of options, choices and opportunities for their own surroundings. This in turn has developed their respect for the talent and work of designers, elevating the status as well as the quality and diversity of the work. Although this democratisation has its downsides in terms of product quality, it has also led to an explosion of talent, creating finer and finer designs that are not only helping to shape the current trend in residential interior design but also paving the way for a future bursting with potential and possibilities.

若要树立现代设计中的典范，我们不应该忘记那些著名的设计师，他们取得了长期的成就，为人们带来了经典的设计和高品质的服务。尼古拉斯·哈斯拉姆、阿尔贝图·品托和雅克·加西亚是少数几个优秀的代表，在住宅的室内设计界一直立于不败之地。他们在更广泛的设计领域也表现得相当出色，以经典风格取代流行元素，以奇特的摆设为空间加分。他们会将18世纪的风格以21世纪的方式表现出来。

最后，什么是不好的设计呢？是不是还有一些糟粕残存在新的设计意识之中？答案当然是肯定的，但这是谁造成的？大批量生产造成了设计的缩水，削减设计费用以降低成本，使产品价格更加平易近人，如果一般民众购买了这些设计平庸的产品，他们又能责怪谁？我们的使命是美化世界，但我们是不是利用这一使命创造了一个怪物？大量与家居设计有关的出版物、书籍和电视节目使公众对家居用品的选择面扩大了，因而人们逐渐尊重设计师的作品，设计师的地位提升了，才会有更多更好的作品诞生。虽然这种大众化对某些产品的质量有些不利，但也促进人才涌现，出现了更精致的设计作品，不仅引领了当今家居室内设计的潮流，还为未来的发展铺平了道路。

Hank M. Chao
ART DIRECTOR
MoHen Design Internationa

赵牧桓
设计总监
牧桓设计+灯光设计研究室

The home living area should be various, unique and reflect the specific elements of the master. To put it in another way, there will not be two identical homes; a home without specialty will not exist. If a home living area happened to be like that, it would remind people of "public living area" such as hotels and executive apartments. No matter how luxury, interesting and charming the public living area is, hundreds of identical homes would make you feel that we are all robots living in the future, without any personality and differentiation. There would not be a coincidence that I encounter an exactly same "me"; therefore every home should be different.

The priority of interior design of home living area should be the consideration of how to reflect the unique characteristics of the master. It is certainly crucial to create a unique, stylish and charming home living area. The design concept can be flamboyance, simplicity, classicality, or serenity. A unique and stylish home living area is the wishes and dreams of the majority of people for their own homes. Here in this book is a collection of the world's most representative, characteristic, and stylish home interior design. Every single project selected has its own specialty and its stylish design process, but they all share something in common – having their own styling and personality.

Designers have the responsibility to assist their clients to locate and realise their own dreams of style, and this should be considered as designers' ethical obligation and social consciousness. We hope that with the harmonious styles in these fabulous cases in the book, we can assist designers to achieve their goals. This will certainly help more people to realise their dreamed future home living areas.

居家环境原本就应该是多样的、个性化的，完全反映主人的特质的。我们也可以说一个住家没理由跟另一个住家长的一模一样，毫无差异性，甚至没有一点个人特性。住家若真变成如此相似，恐怕就给人感觉像是住在酒店或公寓式酒店那种"公共空间"里了，不管那样的装修有多豪华，多有趣，多吸引人，但是上百间住家长的一模一样，只会让人联想到未来世界里的机器人，一点人性化和差异化都没有；再怎么样，全世界都不会有两个完全一样的你和我，每个人家也理所当然地要有所不同。

反映主人与众不同的个性是家居设计里的首要考虑条件，营造具有个人风格和魅力的居所是至关重要的。不管它的概念是华丽、简约、古典、颓废或者是宁静，拥有别具风格的居所绝对是所有人共同的期盼和梦想。在这里，网罗了目前世界上较具代表性、风格迥异及个性突出的家居设计作品。每个个案都有自己的特殊性和设计上的着眼点。但他们的共同性都是具有自己独特味道和风格。

设计师绝对有责任协助每个业主找到他的个人梦想并让梦想变成现实，这也应该是身为设计者不可推诿的道德责任和社会意识。希望透过这些杰出的案例，能够促使设计者更好的实现其设计目标，也协助更多个人更清楚的看见自己所企求的未来居所的模样。

"360" INNOVATIVE LIVING
Hong Kong, China

"360" 创新生活
中国 香港

Johnny Wong and Miho Hirabayashi of FAK3 have created a home in Hong Kong for Joanne Ooi, creative director of Shanghai Tang, adopting the concept of opening multifunctional spaces to a whole new realm. Wong and Hirabayashi have developed an iconic elliptical entertainment cabinet that can rotate at a 360 degree angle, sitting as the central focus in the apartment. On its two longer sides, a set of keyboards and a customised study desk are located; the two shorter sides accommodate storage and a television. Fitted with industrial strength bearings that can support up to two tonnes, the walnut wooden cabinet rotates with the push of just one finger.

Colour plays an important role in 360's design. The kitchen is finished in brilliant cobalt blue. One bathroom is clad in soft pink and white marble while the other sparkles with crystal and jade.

The black shower curtain is custom-made and features pastoral images of Guilin City of China. To maximise the play area, the son's cheerful yellow bedroom features a bunk bed with storage, closet and desk below. In contrast to the large fluid space in the living area, the bedrooms and bathrooms are orthogonal and strictly delineated.

FAK3公司的强尼·黄和美穗平林为"上海唐"的创意总监乔安妮·奥依设计了她在香港的家。他们采用开放式多功能空间的理念，营造出一个全新的境界。黄和平林设计了一个经典的椭圆形影音橱柜。它可以360度旋转，是整个公寓最引人注目的焦点。影音柜较宽的两边，分别装着键盘和特别定制的书桌；较窄的两侧分别是收纳空间和电视机位。橱柜装有工业轴承，可以承受多达两吨的重量。要推动这座核桃木的旋转橱柜，只需一根手指的力量。它的轴承中心偏移，橱柜转动时，其中放置的物品及其架构似乎让房子的空间结构都产生了变化，柜子每个分区的特色都展现了出来。

在360的设计中，色彩发挥了很大的作用。厨房是夺目的钴蓝色。浴室穿着淡粉色和白色的大理石外衣，其中点缀着闪烁的水晶和玉石。黑色浴帘是定制的，上面画着桂林山水图。为了将游戏空间最大化，在主人儿子的卧室，功能分布呈上下结构，步入上层床铺的楼梯也可用于储物，壁橱和书桌位于下部，色调选用明快的黄色。与起居空间的随意性相反，卧室和浴室都设计得中规中矩，界限分明。

PROJECT
"360" Innovative living
DESIGNER
Johnny Wong and Miho Hirabayashi
COMPLETION
2007
PHOTOGRAPHER
FAK3

项目
"360" 创新生活
设计师
强尼·黄，美穗平林
完工时间
2007
摄影师
FAK3

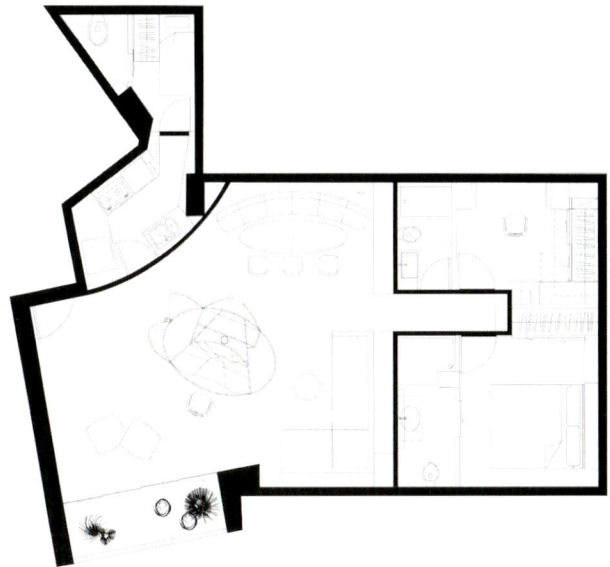

ALMA ROAD RESIDENCE
Melbourne, Australia

阿尔玛住宅
澳大利亚 墨尔本

This home wears many guises – a formal entertainer, a family incubator and a creative hub. In contrast to the formality of the Victorian original, it is warm and liveable; it embraces kids and dogs, a busy lifestyle and a constant traffic that goes with it. None of the spaces are reserved for "good"; for example, each room could be used for a range of purposes because the design is versatile and accommodating. In addition, responding to disparate functional demands, a canvas is provided for an extensive contemporary art collection and an eclectic mix of furniture and favourite objects.

Quite simply, the main idea was the brief – to create a family home out of an 1880s' Victorian mansion that had been converted into eight bedsit apartments in the 1930s. The aim was to awaken the splendour of the original and to inject it to the aesthetic and functional realities of the 21st century.

这栋住宅兼具着多重身份：既是正式的款待者，又是家庭的维系者，还是创意的汇集地。不同于其原本维多利亚风格的拘谨俗套，如今它已变得温馨舒适。房子为孩子和狗提供了舒服的居住空间，即使你每天进进出出，忙碌不堪，它始终为你提供着便利。房子的设计灵活而周到，这里没有仅供观赏的空间，每个房间都有很多的用途。对于不同的功能要求，房子里设立了一顶帐篷，里面有大量的当代艺术收藏品，还有许多的家具和信手拈来的材料。

房子的总体设计理念很简单：创建一个家庭住宅，跳脱出19世纪80年代维多利亚风格建筑的框架，改建为20世纪30年代有八个卧室兼起居室的公寓。旨在让旧居焕发光彩，使之成为兼顾美学效果和功能实用性的21世纪新式住宅。

PROJECT
Alma Road residence
DESIGNER
Hecker Phelan & Guthrie in association with BURO Architects and Nonda Katsalidis
COMPLETION
2005
PHOTOGRAPHER
Shannon McGrath

项目
阿尔玛住宅
设计师
Hecker Phelan & Guthrie in association with BURO Architects and Nonda Katsalidis
完工时间
2005
摄影师
Shannon McGrath

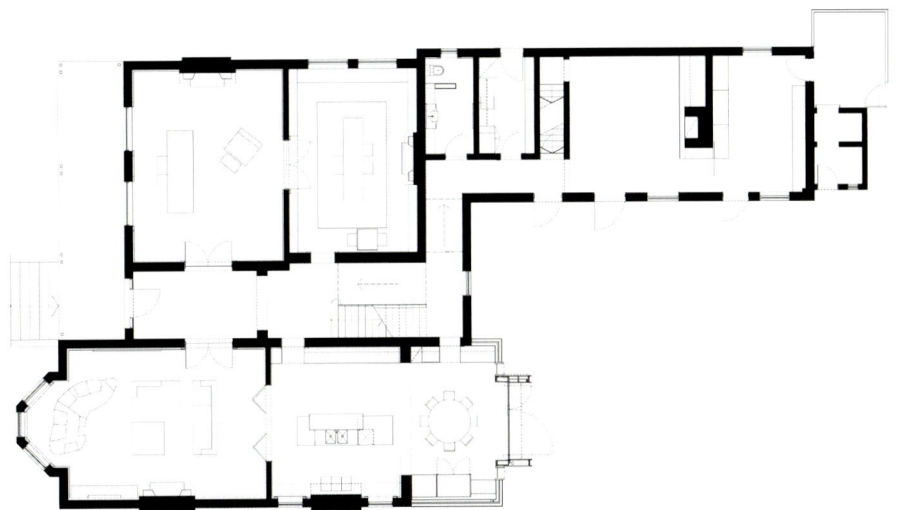

APARTMENT S
Stuttgart, Germany

S公寓
德国 斯图加特

The focus of this residence was to create a continuous open space while maintaining the connection of livable spaces through the use of materials, lighting, and ceiling elements. Once the walls were removed, the home opened itself up from one end to the other. This allowed for a functional progression and an abundance of natural light to flood the space.

Piercing through the entire length of the home, a ribbed structure, doubling as a light fixture, threads each space terminating in the dropped ceiling above the fireplace/ entertainment wall. This wall features a sliding wood panel that functions as a curtain allowing for control of what is seen. A connection of the main spaces is achieved through the path that was laid in the floor by contrasting types of wood.

这个住宅在保持适宜居住的前提下，通过原材料、照明设施以及天花板的运用，打造一个具有连贯性的开放空间。一旦移开墙壁，住宅就成为一个完全开放的空间，这样就大大提升了空间的功能性和采光效果。

一个横贯整个空间的肋架结构穿过每一个空间直达壁炉/视听设备陈列柜上方狭小的吊顶。与此同时，它还兼扮着灯光装置的角色。以一块滑动的木板为墙，像窗帘一样起到遮挡视线的作用。利用反差极大的木材铺设而成的"地板路"保持了主空间的连贯性。

PROJECT
Apartment S
DESIGNER
ippolito fleitz group
COMPLETION
2007
PHOTOGRAPHER
Zooey Braun

项目
S公寓
设计师
ippolito fleitz group
完工时间
2007
摄影师
佐伊·布劳恩

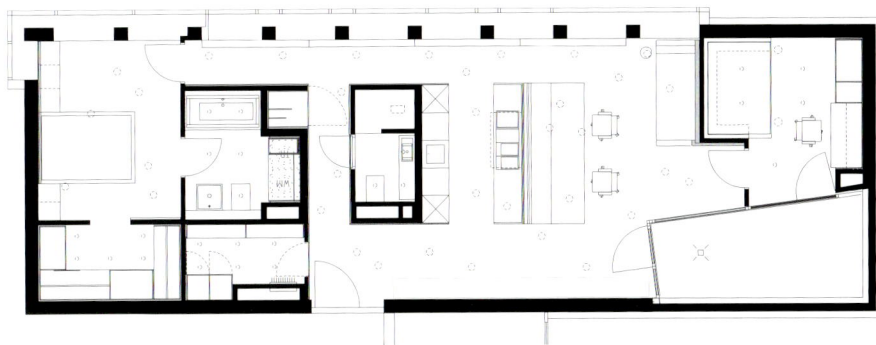

ANGLER'S BAY SHOW FLAT
Hongkong, China

海云轩
中国 香港

Mixed with minimalism and cyber-concept, white colour scheme is applied on this 661-square-foot apartment among the walls, the tiles and the dinning table. The mirror reflects not only the lights but also the spatial relationship.

To run the idea of "cyber", the designer takes the advantage of material such as glass and plastic. The dinning chairs, the ceiling-mounted light fixture, and the coffee table are all transparent and futuristic. Besides, the black carpet and the grey sofa create a visual balance with the white enclosure wall. The 36-square-foot balcony provides a gorgeous view of Ma Wan Channel and Tsing Ma Bridge.

The original two bedrooms have been combined together as a 160-square-foot bedroom, which is separated from the living area by a dynamic element – a full height rotatable glossy champagne cabinet. With the surface-mounted plasma TV, the dwellers can enjoy their favourite shows either in the bedroom or in the living room.

项目面积661平方尺，单位混合简约和cyber的概念，走进公寓，便可看见全屋以白色为主，白色的墙、地砖及餐桌，素雅简洁。餐桌旁放上一面镜子，不但使光线充足，更加强空间感。

为营造cyber效果，设计师利用了玻璃和塑料材料的优势，餐桌的4张椅子、吊灯和茶几都是透明的，具现代感和前卫感。大厅放置灰色的沙发和黑色地毯，与白色墙体配衬得宜。站在大厅外连的36平方尺的露台可以眺望青马大桥、汀九桥和汲水门大桥美景，景观开阔。

卧室内以深色布置。原为两个房间设计，现打通为一房。设计师将客厅与睡房的墙拆除，以一个金色可旋转的落地柜分隔，落地玻璃背后安装电视，可把电视转向睡房，设计新颖又照顾居住者的需要，亦令屋内空间看起来更宽阔。此外，书桌设于窗台上，可以一边工作，一边眺望马湾美景。

PROJECT
Angler's Bay Show flat
DESIGNER
PTang Studio Ltd

COMPLETION
2006
PHOTOGRAPHER
Joe Murray

项目
海云轩
设计师
PTang Studio Ltd

完工时间
2006
摄影师
Joe Murray

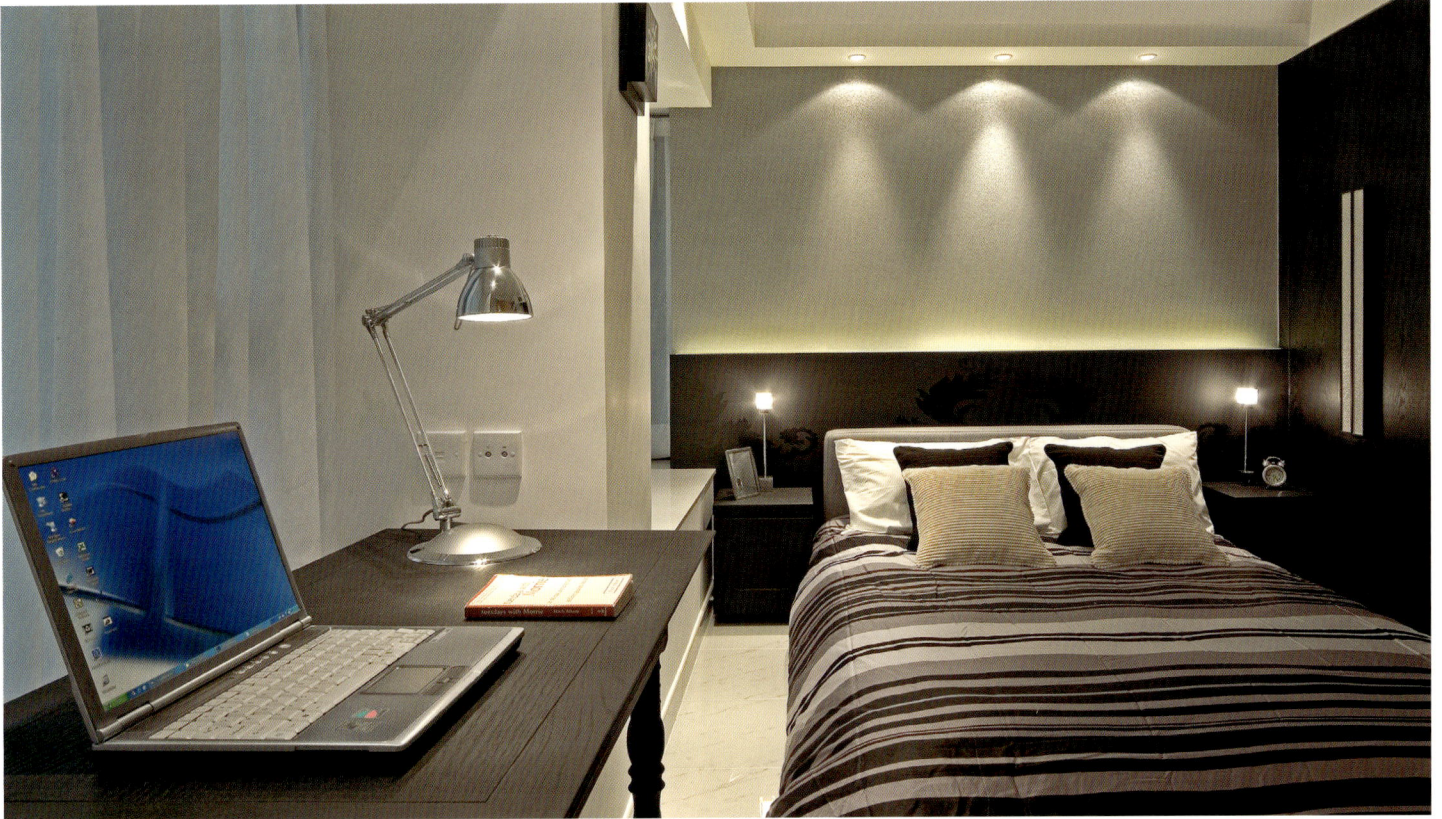

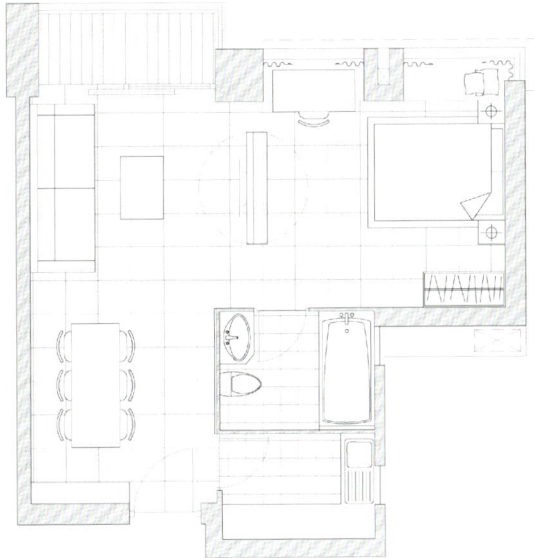

BEIJING CHATEAU TOWERS
Beijing, China

北京公馆
中国 北京

Representative Party Venue, Company Guest House, Luxury Private Apartment – these are the various requirements for the high-end condominium developments in the dynamic real estate market of the Chinese capital. As a showroom for a project of large-scale apartments, Graft developed a flexible layout, dominated by a dynamic sculpture which separates the private rooms from the public reception areas.

By semitransparent space dividers carved in association with Chinese lattice patterns, the apartment can be shifted within seconds between an all open entertainment facility and an intimate family home.

有代表性的派对大厅、商务客房、 豪华的私人空间。这些都是中国首都——北京的动态的房地产市场对高端公寓的要求。Graft使用了一种灵活的布局方式来打造这个大规模公寓的样板间。一尊可移动的雕塑将私人空间同公共的接待空间分隔开来。

半透明的空间隔断、刻有花纹的中式格子框架使公寓能够在数秒间完成从全开放的娱乐空间到私人的家庭空间的自由转换。

PROJECT
Beijing Chateau Towers
DESIGNER
Graft
COMPLETION
2008
PHOTOGRAPHER
Fang Zhenning

项目
北京公馆
设计师
Graft
完工时间
2008
摄影师
方振宁

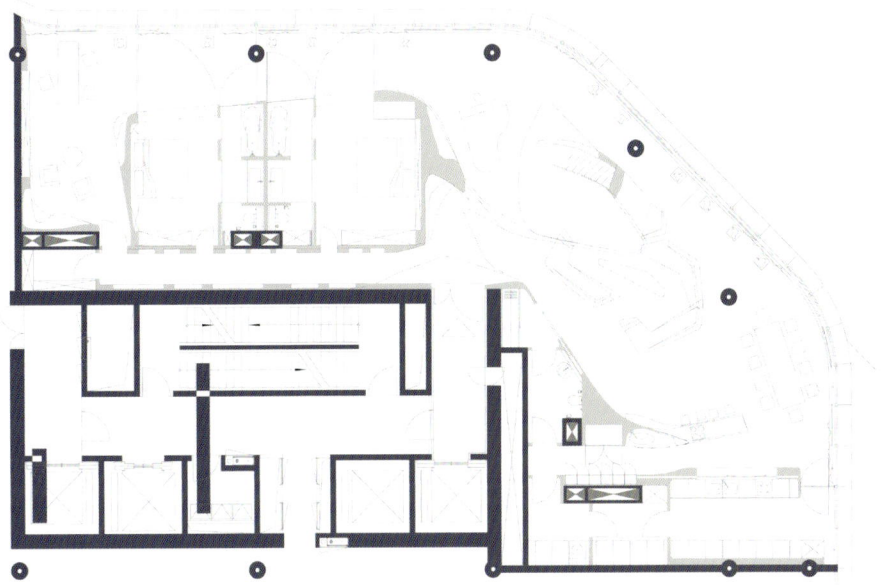

BOHEMIAN EMBASSY BEDROOM MODEL SUITE
Toronto, Canada

波西米亚大使馆套房样板间
加拿大 多伦多

This generously-proportioned model suite is designed in a sophisticated urban style, with a serene palette of warm grey and driftwood providing an elegant backdrop for artwork. Stylish modern fittings and furnishings are balanced with unexpected traditional details and accessories that suggest a creative spirit and an artistic lifestyle.

The main living and dining groups decorated with classic contemporary furnishings are defined by a dense, grey plush wool area rug and lit by a sparkling pendant fixture of small mirrored discs. The combination of modern and traditional elements continues into the white marble-clad master bath, and similar fittings are used in the guest bath. In the guest bedroom, an oversized, ornately-carved and grey-painted table contrasts with the clean-lined quilted linen headboard and the modern rug.

宁静柔和的灰色调和雅致的浮木艺术品背景墙使这个样板套房具有了典雅的都市风格。极具格调的现代家具和陈设，与蕴含构成创意城市新元素的创造精神和艺术生活方式的传统配饰完美融合。

起居和就餐区里典雅的现代家具被一块灰色的长绒羊毛地毯衬托。一盏圆形金属亮片吊灯点亮了整个空间。白色大理石装点的主浴间也延续了传统与现代相融合的特点。客浴间也用了相似的设备和器具。在客卧里，一张巨大而华丽的灰色刻花桌，与线条简洁的亚麻布床头板同时尚的小地毯形成鲜明的对比。

PROJECT
Bohemian Embassy Bedroom Model Suite
DESIGNER
II BY IV Design Associates Inc.
COMPLETION
2006
PHOTOGRAPHER
David Whittaker

项目
波西米亚大使馆套房样板间
设计师
II BY IV Design Associates Inc.
完工时间
2006
摄影师
David Whittaker

CAMPOS ELISEOS APARTMENT
Mexico city, Mexico

伊莉斯草原小屋
墨西哥 墨西哥城

JSA was commissioned to make the interior design of a 6,674-square-foot apartment, in a sober and elegant style. Very spacious open spaces for mixed uses were linked to create common areas for leisure activities, furnished and decorated in a warm style to recapture the family scale.

Richly-textured materials were combined with classic large-format marble and different woods, thus achieving the feeling the designer and the client were seeking for.

一种端庄文雅的风格在这个面积为6674平方英尺的家居项目中体现出来。可多样使用的开阔空间与可进行休闲活动的创意区域相连接，它的装饰和家具都营造了一种温暖的家庭感。

大量地运用了有织纹的材料、大块的高级大理石和不同风格、强烈线条感的木制材料，成功地营造了客户所追求的感觉。

PROJECT
Campos Eliseos Apartment
DESIGNER
JSA
COMPLETION
2007
PHOTOGRAPHER
Paúl Rivera - Archphoto

项目
伊莉斯草原小屋
设计师
JSA
完工时间
2007
摄影师
Paúl Rivera – Archphoto

GREEN PENTHOUSE LOFT
New York, USA

绿色小棚屋
美国 纽约

The design strategy was to separate the "primary" and "support" functions. The "primary" spaces are aligned and sequenced along the perimeter window wall, which acts as a porous plane with the living room/kitchen/dining room/master bedroom on the inside and the terrace on the outside. The visual perception of the perimeter window wall on the inside is a sequence of layered spaces, allowing the eye to traverse the length of the loft, through the apertures and transparency of the furniture-like kitchen cabinetry as well as the glazed opening in the master bedroom wall.

The "support" spaces are arranged along the entry wall as closets, the powder room/master bathroom/walk-in closet, defining also the spatial perimeter of the "primary" spaces. In between the "primary" and "support" spaces a continuous diagonal line of sight is preserved to emphasise the spatial depth of the interior in parallel with the expansive view from the terrace.

设计策略是分割并分别构建"主要"及"次要"功能。"主要"空间沿玻璃幕墙一字排开，依次是起居室、厨房、餐厅和主卧室，由透气性好的面板相隔；外面是阳台。透过玻璃幕墙看到顺序感极强的空间，视线能看到整个楼面，从厨房透明的橱柜和缝隙之间，甚至能看到主卧室的开放式釉面墙壁。

"次要"空间靠着入户墙，依次是壁橱、化妆室、主卫生间和步入式衣帽间，这些房间的间隔同样极具空间感。在"主要"和"次要"空间中，按对角线方向能够不间断的望到对面，加强了内部空间的深度，并从阳台的角度开阔了视野。

PROJECT
Green Penthouse Loft
DESIGNER
Cha & Innerhofer Architecture + Design
COMPLETION
2003
PHOTOGRAPHER
Dao Lou Zha

项目
绿色小棚屋
设计师
Cha & Innerhofer Architecture + Design
完工时间
2003
摄影师
Dao Lou Zha

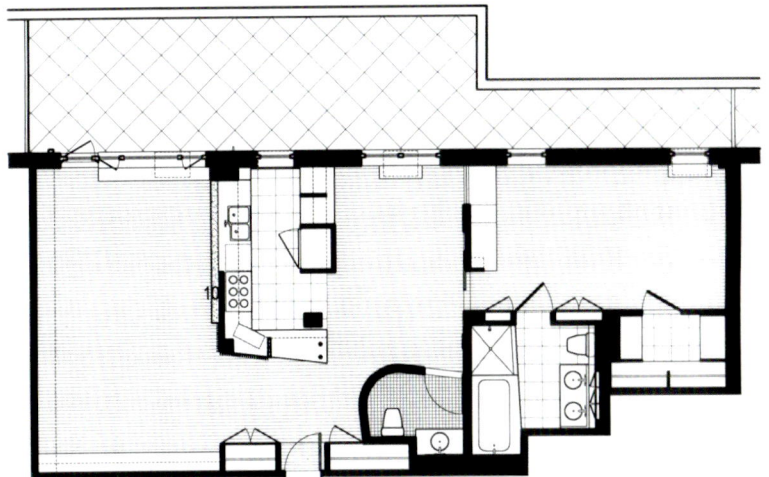

HARDWICK TURNBULL BEACH HOUSE
Sydney, Australia

哈德维克通布尔海景房
澳大利亚 悉尼

This original 1960's home was already in good condition but needed updating. The designers reworked the original 1960's style with mid-century modern iconic furniture of the likes of Eames and Jacobsen. These were mixed with modern furniture from Minotti and new custom-made carpets.

The colour palette was inspired by the surroundings of the house. Since the house is located by the ocean, the designers chose to use materials that refer to the sand, water, timbers and fauna of the local Palm Beach area.

The outdoor lounge area is the reverse of this, where Philip Stark's outdoor sofa was used to create an outdoor room. The room is walled in by the use of bamboo and the floor is paved with natural local sandstone.

这个海景房始建于20世纪60年代，设施已十分齐备，但仍需更新。设计师选用Eames和Jacobsen的中世纪经典家具，对这幢60年代风格的建筑进行重新改造。同时使用了Minotti的现代家具和定制的新式地毯。

色调的灵感来源于度假村的周边景观。因为靠海，设计师选用了沙子、水、木材和棕榈等与当地景物相一致的材料。还有蓝色的装饰品和漆器，黄色的织物和仿天然石的庭院。

我们用菲利普斯塔克的户外沙发营造出室内露天的效果，颠覆了户外休闲区的概念。房间内用竹子做成墙壁，地板上则铺满当地的天然砂岩。

PROJECT
Hardwick Turnbull Beach House
DESIGNER
Greg Natale Design
COMPLETION
2004
PHOTOGRAPHER
Anson Smart

项目
哈德维克通布尔海景房
设计师
Greg Natale Design
完工时间
2004
摄影师
Anson Smart

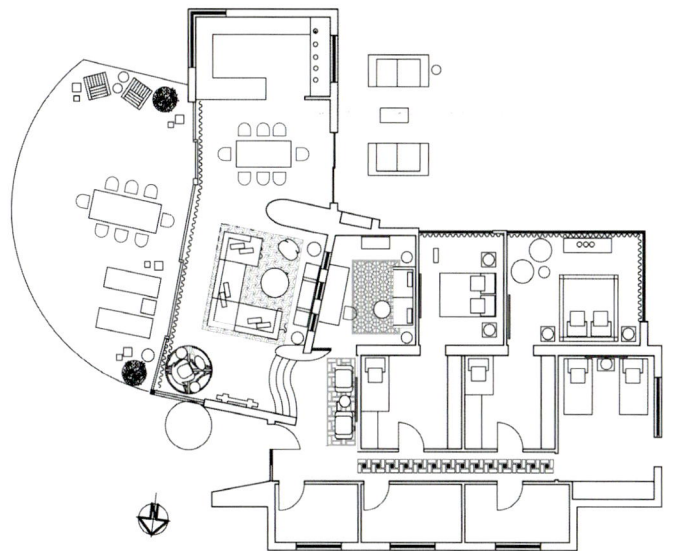

HSIEH'S RESIDENT
Taiwan, China

谢公馆
中国 台湾

Oriental style has made frequent appearances all around the world and this style is actually not outstanding anymore. In order to achieve the client's requirements, the designers get the theme "oriental fusion".

"Oriental fusion" here aims at presenting the Asian art and craft style in a modern way and also mix-and-match with Western ornaments, i.e. crystal and modern metal sculpture. The design needs not to be historically accordant, but everything must be tightened up together harmoniously and in the right mood.

The living and the dining areas are separated with a semi-transparent screen. The shape of this custom-made screen is inspired by Asian temples and laminated with gold leaves. The design of the chandelier in the dining area is inspired by the pavilion of the Wang Shi Yuan in Jiang Nan. The furniture is selected very carefully in terms of design and upholstery colour, and also with environmental consideration.

东方风格在世界各地都十分常见。事实上，这种风格已有些过时。为了满足客户的需求，设计师以"东方融合"为整个空间的设计主题。

"东方融合"的目的在于用现代的方式以及与西方装饰艺术（如水晶、现代金属雕塑）混搭的方法，来展现亚洲工艺品。设计不必完全遵循历史。但是，所有东西不仅要和谐地相融合，而且要与整体的风格相协调。

一扇半透明的玻璃屏风将起居室和就餐区隔离开来。定制的玻璃屏风造型独特，其灵感来自于亚洲的庙宇。另外，屏风上还点缀有精致的金色树叶。就餐区的枝形吊灯，灵感来自江南网师园的亭阁。家具都是为了配合设计师主题精心挑选而来。室内家居的选择也充分考虑到了与环境融合的问题。

PROJECT
HSIEH's Resident
DESIGNER
Tien fun Interior Planning Co., Ltd.
COMPLETION
2007
PHOTOGRAPHER
Shou Shan LAI

项目
谢公馆
设计师
天坊室内规划有限公司
完工时间
2007
摄影师
Shou Shan LAI

KENSINGTON HOUSE
Kensington, UK

凯盛顿公寓
英国 肯辛顿

SHH took on the newly-built Kensington House project with a commission to create a complete interior scheme for the six-storey, 1,114-square-metre house. The client's brief was simple – the house had to suit the needs of an international couple with a large family.

The newly-built house is a contemporary Georgian terrace house, set within a classic square, combining all the conveniences of modern living with state-of-the-art AV and comfort cooling, without compromising space and traditional proportions. The property is comprised of a very generously-proportioned reception room, a large formal dining room, a family room, a cinema and a pool room, all located on the lower three floors. The entire ground floor is occupied by the master suite, with the remaining two floors taken up by five further bedrooms and en suite bathrooms.

英国建筑师和设计师承担了重建肯辛顿公寓的任务。根据客户的要求，设计师们要为1114平方米的六层建筑制定一个完整的设计策划。客户的要求很简单，大厦必须能满足来自世界各地的、子女众多的夫妇的要求。至于大厦的外观，白色而且时尚即可。

这栋新建的当代公寓，呈现出一个带有乔治亚州式平台的房型，内部设有一个传统的方形客厅，在之中将高科技影音设备与舒适便捷的现代生活相结合，设计并没有采用传统均衡空间的做法。

整栋大厦由宽敞的会客厅、餐厅、家庭活动室、影音室和桌球室组成，这些房间都集中在大厦的一至三层，其中，整个二层都是主套间，一层和三层除了主套间之外，还有5间卧房和浴室套房。

PROJECT
KENSINGTON HOUSE
DESIGNER
SHH
COMPLETION
2007
PHOTOGRAPHER
James Silverman

项目
凯盛顿公寓
设计师
SHH
完工时间
2007
摄影师
詹姆斯·西尔弗曼

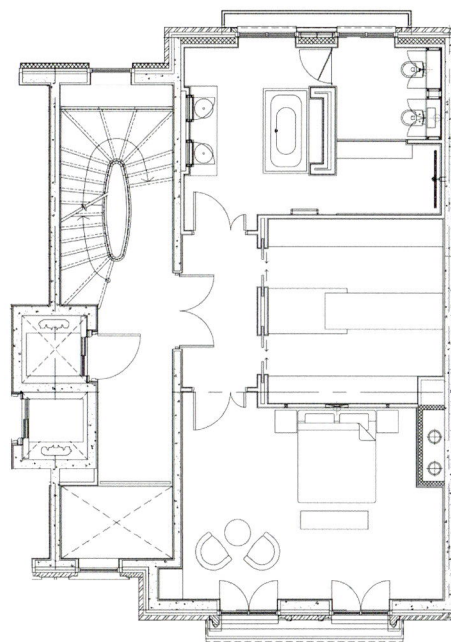

KENSINGTON TOWNHOUSE
London, UK

凯盛顿联排别墅
英国 伦敦

Internally, the house was stripped down to a bare shell and room layouts were opened up to make a more practicable environment for a modern family. With such a location, while some aspects of material and colour choices were conservative, the use of dark grey porcelain floors for the raised ground sitting room was a departure from the wood floors generally found in such houses. The dark colour was offset by a glossy finish to reflect the light. A minimal black stone fireplace makes a dramatic statement against the white walls and enhances the flames at night.

A mix of Asian artifacts, African and contemporary Italian furniture was used throughout, the dark woods of the furniture working with the black and grey tones of the sitting room. A specific request was that the baby room also benefits from good furniture and design; a cherry prototype cabinet, an oak "tots" dining table and chairs work well with the large American lacquered cot.

房间采用完全开放性的布局增强了空间感，内部没有任何的隔挡。由于环境的特点，材料和颜色的选择偏于保守，没有选择传统的木制地板，而采用了深灰色的地砖，是这间起居室的一个亮点。光泽感极强的罩面漆具有良好的反光性，从而弥补了暗色调带来的压抑感。一个迷你的黑色石质壁炉与白色墙面形成强烈的对比，使夜色中的火焰更加鲜明。

来自亚洲的古董和来自非洲、意大利的家具，充斥着整个空间。黑色的木制家具，不但与起居室里黑灰色调相呼应，更衬托了厨房里的橡木椅架。婴儿房里的家具和设计也别具匠心。樱桃木的立柜、橡木餐桌椅与美国儿童木漆床和谐搭配。

PROJECT
Kensington Townhouse
DESIGNER
Kinari design
COMPLETION
2006

项目
凯盛顿联排别墅
设计师
Kinari design
完工时间
2006

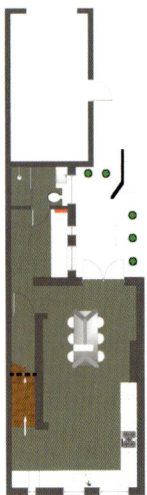
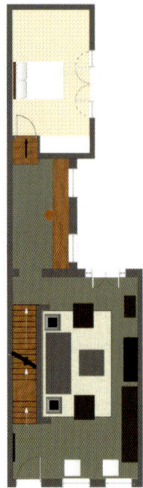

LAS LOMAS
Madrid, Spai

洛马斯洋房
西班牙 马德里

In Las Lomas neighbourhood, a few kilometres from the city of Madrid, this detached house is located. The main characters of the home are its pieces of art; the owners have one of the most important Chinese art collections in Spain.

The furniture has been designed to allow each piece of the collection as unique as it is, so the decoration gives way to the collection, but at the same time, the decoration is as excellent as each of the collection pieces. The materials used are wengué and silver paper.

在洛马斯社区，距喧闹的马德里市数公里之遥，遥望山脉时，就能看见这幢独立洋房。房子的主要特征是大量的艺术品，它的主人在西班牙拥有一件极具价值的中国艺术收藏品。

家具的每一处设计都独树一帜，装饰手法抢去了藏品的风头，看上去却也和谐一致。所用的材料包括黑铁木、银纸、奶油色和质地优良的织物。

PROJECT
LAS LOMAS
DESIGNER
Estudio Interiors
COMPLETION
2006

项目
洛马斯洋房
设计师
Estudio Interiors
完工时间
2006

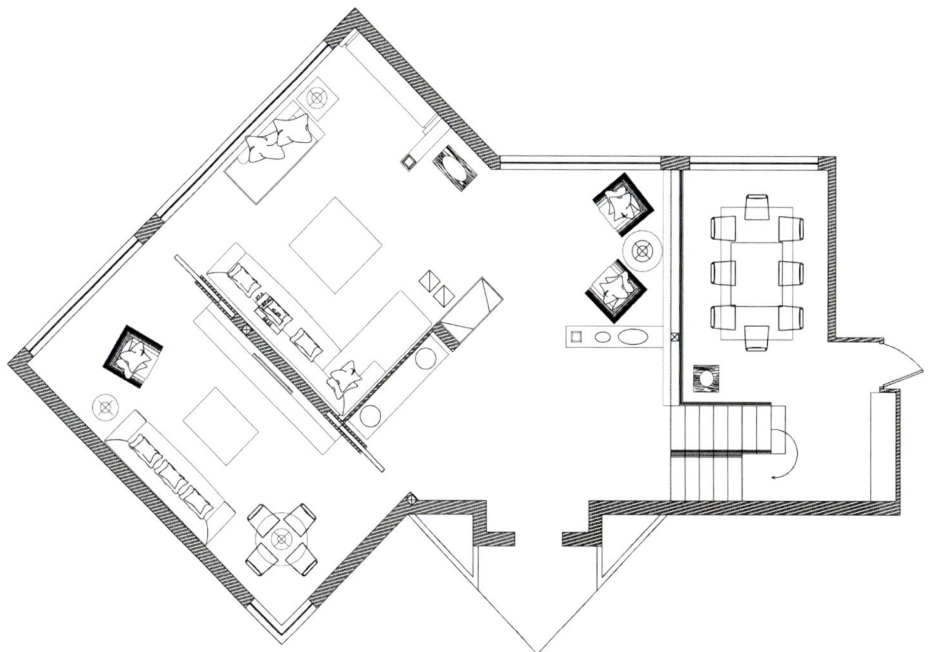

LOFT IN MADRID
Madrid, Spai

马德里公寓
西班牙 马德里

This cosmopolitan duplex apartment is located in the centre of Madrid. It exudes sober and elegant, glamorous and spontaneous feeling. The spaces reflect every character of the owner, his way of living, his philosophy of life, etc. Gleaming finishes in woods, leathers, and coverings make it a fashionable place to live in.

The flowers make the room much fresh, while the art murals offer an elegant atmosphere. It is not only a home; it is more than a hotel or a gallery. It was voted the most chic apartment in Madrid by readers of Style magazine.

这座国际双层公寓坐落在马德里的市中心。它给人们的感觉就像是一个老成持重、风度翩翩、富有魅力、自然随性的王者，正如他的拥有者一样。公寓反映了主人性格的各个方面，他的生活方式和处世哲学。室内装饰采用了木材、皮革和地毯，把它装点成一个时尚的住所。

鲜花使房间变得更加清新，艺术壁画使室内氛围如此高雅。它不仅是一个家，更超越了酒店或画廊。它被《时尚》杂志的读者评为马德里最别致的公寓。

PROJECT
LOFT IN MADRID
DESIGNER
Estudio interiors
COMPLETION
2006

项目
马德里公寓
设计师
Estudio interiors
完工时间
2006

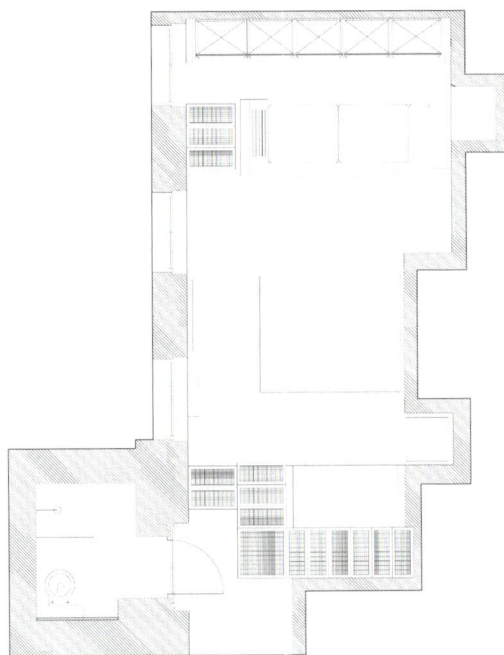

LA ROSSA
Shenzhen, China

映岸红
中国 深圳

Stylish red and pure white are the skeleton of the design direction. Flower patterns around the home form a very warm atmosphere for this three-bedroom sea-view unit. A red TV cabinet acts as a red carpet, to serve as a foil to a contrasting graphical tree wall, making this square-like living room a romantic place.

A tree-like bookshelf in the study is not only functional, but also becomes the focus of the unit. Its grey square boxes and the big red translucent flowers on the door create a dramatic contrast.

In the master bedroom, the bedding, the wallpaper and the wardrobe doors are "covered" by "flowers", which continues the theme of "flower" in the whole design.

设计师把时髦的红色、纯洁的白色和妙曼的花卉图案作为设计方向，把这个3房海景人家，创造成与众不同的现代舒适居室风格。客厅运用花蕾图案的墙纸；与长形的鲜红色电视柜相连接，配衬一张鲜红的地毯，让这个空间充满浪漫的气息。

形似一棵树的书架不仅实现了其功能性的要求，也是房间的一个亮点。灰色的四方盒子与门上巨大的半透明红色花朵图案形成了鲜明的对比。

主人套房空间宽敞，偌大的主人床，以银色花为图案的墙纸，加上白色玻璃门落地衣柜的磨砂花纹，令房间尽显高雅瑰丽，这些元素都是整个设计作品"花"主题的延续。

PROJECT
La Rossa
DESIGNER
PTang Studio Ltd
COMPLETION
2007
PHOTOGRAPHER
Brian Ip

项目
映岸红
设计师
PTang Studio Ltd
完工时间
2007
摄影师
Brian Ip

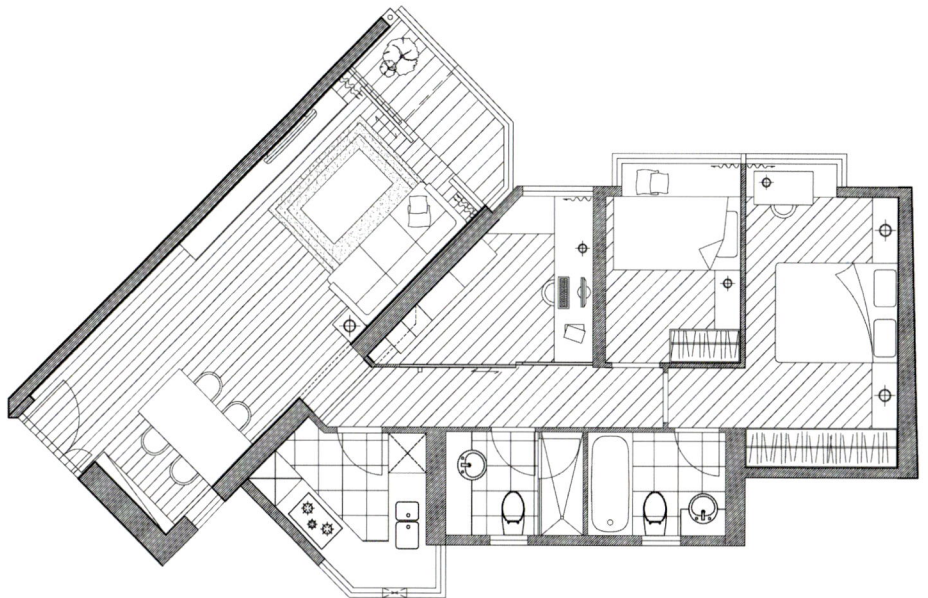

MILLENNIUM TOWER LOFT
Seattle, USA

千年塔阁楼
美国 西雅图

The goal is to combine wide open spaces for entertainment, with enough privacy for a small family with a child. The main floor layout is designed to transform from a single-volume entertainment space to distinct kitchen, dining room, entrance, and guest mini-apartment with discreet built-in closets; the extra deep sofa combines with the ottoman to create a full bed, and the powder room contains a hidden shower tucked away behind angled privacy glass panels. There is a flexible office space on the ground floor that can be used for different family activities and can, in the future, be transformed into a bedroom if the need arises.

Humans like variety. Single look and feel, no matter how good, cannot work perfectly in all situations. Glass in a variety of colours and patterns take on multiple functions. Transparent, muted with bronze tint, layered with graphical elements used on vertical and horizontal surfaces as part of the floor, walls, doors and the ceiling – glass becomes the main theme in layered transparency of the space. The aluminum-framed floating glass floor of the office space, as well as the movable partitions, helps to make different impressions and change perspectives.

这座阁楼既有供人娱乐的开放空间，又兼顾了一个三口之家所需的充分的私密性，还是舒适的待客之所。一楼不再是单一的娱乐空间，还分隔出厨房、饭厅、玄关和客用迷你公寓，内设步入式衣帽间，独立的沙发，加上脚凳它就能变成一张床，盥洗室里有调光玻璃板搭成的隐形浴室。二楼空间可以灵活使用，既能举行各种家庭活动，将来有需要时，还能作为第三个房间。

人类偏爱多样化。单一的外观和感觉，不管它有多好，也不能在所有情形下有完美的表现，比如白天与黑夜，细雨绵绵与晴空万里，一家人聚在一起时的热闹与夜晚的安静。颜色和形状各异的玻璃有着不同的作用。透明色和柔和的青铜色玻璃组成生动的图案镶嵌在地板、墙面、门和天花板上——玻璃成为创造空间层次感和透明度的主角。书房的铝框玻璃面地板起到动态分割空间的作用，营造出不一样的心境和视觉效果。

PROJECT
Millennium Tower Loft
DESIGNER
Julia Sandetskaya, Polina Zaika
COMPLETION
2007
PHOTOGRAPHER
Andersen Ross, Alex Mogilevsky

项目
千年塔阁楼
设计师
朱莉娅·桑德斯科娃，波莉娜·扎依卡
完工时间
2007
摄影师
安德生·罗斯，亚历克斯·莫基列维斯基

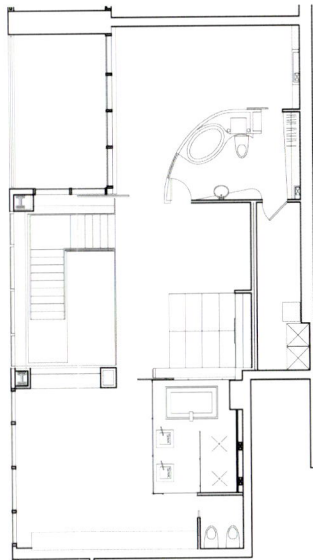

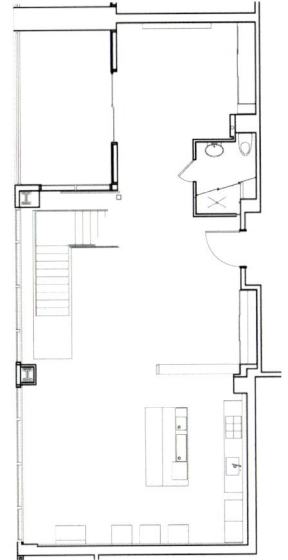

MANGROVE WEST COAST
Shenzhen, China

红树西岸
中国 深圳

To give a profound impression to all visitors, the designer chose "flower" as the theme of this flat. Flower patterns can be found everywhere – on walls, glasses, accessories, etc. The fresh feeling is reinforced by the bright colours of red and white. Red cushions, chairs, and artworks in the flat become the focus.

Flower patterns create shadows and add layers on the plain wall. Besides, bird patterns along the corridor act as a guide to the bedrooms. The "flower" continues in the bedrooms on beddings, artworks and wall graphics, creating a refreshing and warm atmosphere.

设计师利用了花及大自然作为设计主题，给所有的到访者留下了深刻的印象。客厅墙上的立体花卉图案，为白色的墙添加了层次感，而客厅的图案花玻璃镜，视觉上把客厅的深度增加，而运用的黑色，也成为了本项目的焦点。

花卉的平面造型在墙壁上，制造出了带有阴影的层次感。除此之外，走廊墙壁上布满了燕子的图案，彷佛百鸟归巢的景象，带领户主到两个卧室。卧室空间宽敞，贯彻设计主题，用了较平面化的图案，但用上红色的家具作点缀，而卧铺均使用颜色鲜艳的图案，令整个卧间充满清新的气息。

PROJECT
Mangrove West Coast
DESIGNER
PTang Studio Ltd
COMPLETION
2007
PHOTOGRAPHER
Brian Ip

项目
红树西岸
设计师
PTang Studio Ltd
完工时间
2007
摄影师
Brian Ip

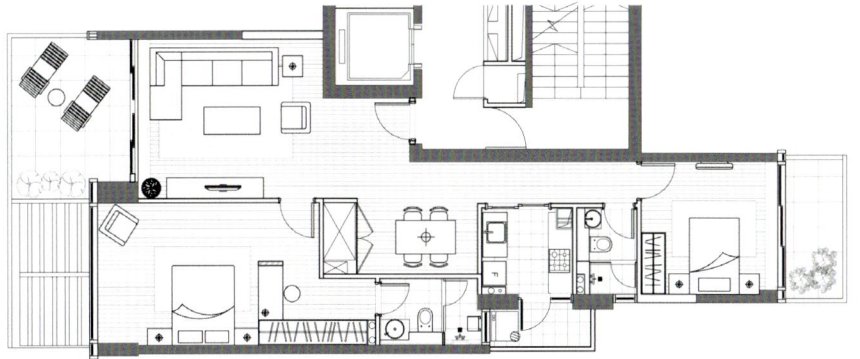

METROPOLITAN CHIC
Shanghai, China

美丽都市
中国 上海

For this project, the use of material is much controlled. The spaces are clearly separated, with the hallway dividing the whole space into two sides. The living room faces the kitchen while the master bedroom faces the small guest room, which is also used as study room. Steel bars are embedded in the wall, making a strong contrast with materials and details of the design, mainly for the purpose to hide the CD cases inside.

The frequent use of mirrors widens this small and compact residential apartment; you can see from the cupboard, the wall back of the sofa, the kitchen and the back wall of the master bed. The brown colour and the mirrors make a good combination. Light yellow and light brown make the space look simple. The materials are also affordable and easy to get. It is suitable to use these kinds of materials for ordinary households. Simple luxury is the right expression for the living character of such a group of people living in Metropolitan Chic.

这个设计的设计语言和材料都是比较收敛的，从某个观点来看也算是比较洁净不张扬的。空间的主轴划分得利落清楚，入口玄关划开了左右两边的空间，客厅对应餐厅、主卧对应小客房兼书房。多余的装饰线条尽量都被舍弃，只留下电视的主墙面的分隔面板嵌进了小的不锈钢条做了点材料上的对比和细腻的设计，主要是为了隐藏在里面的CD储藏柜。大的量体空间利用天花板上的线条、凹凸和光线界定各个空间的地界。

大量的运用镜子让这个原本其实很小、很紧凑的居所在视觉上宽敞了许多，从入口的鞋柜、沙发背景墙、厨房到主卧背墙，茶色和明镜交错运用。米色和浅咖啡色调让空间显得更明快，也对应了整体线条的利落干净。所有的建材都是便宜随手可得的简单材料，用这样的主材用料去表现这种平民式的贵族应该会比较恰当，也更符合美丽都市这样的主题族群的生活特质——简单的奢华。

PROJECT
Metropolitan Chic
DESIGNER
Hank M. Chao / MoHen Design International
COMPLETION
2006
PHOTOGRAPHER
MoHen Design International/Maoder Chou

项目
美丽都市
设计师
牧桓建筑+灯光设计
完工时间
2006
摄影师
MoHen Design International/周宇贤

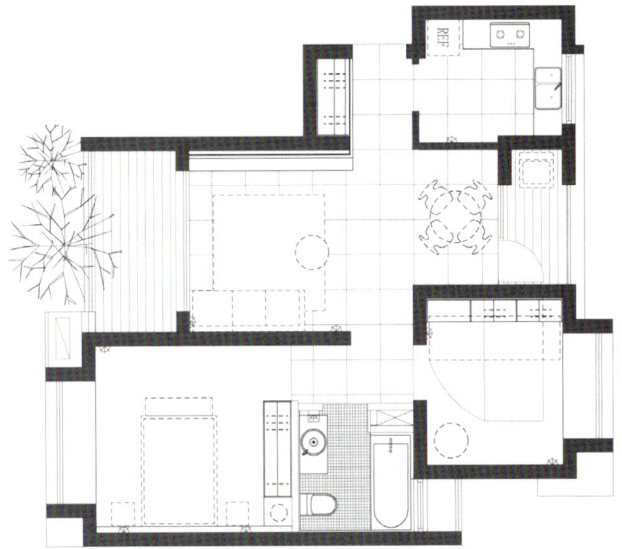

NATURAL ORGANIC SPACE
Wuxi, China

自然有机空间
中国 无锡

The central body line and the horizontal active line on the plan separate the ground floor into several main areas, on one half the living room and on the other half the bedroom. The designers do not want any obstacles to cut the view in the living room and try to expand the vision to its maximal size. The doorway extends to the end of the hallway in order to be combined with the dining area and to be connected with the kitchen.

The lower side of the television wall in the living room is left with an open hole, because it is intended to be connected with the study area in the master bedroom as a breathing tunnel in the visual prospect. The walls in the master bedroom are all designed open; it is connected with the shower room in the bathroom. There is no wall in between in the whole master bedroom; therefore, the fitting room and the study room look much clear in any view. The design of the basement is in accordance with the same style as the ground floor.

平面图上的中轴线和横向动线清楚地把一楼的平面切割成几个主要区块，一边是客厅，一边是卧房。为了把空间视觉的张力做到最大，设计师不想在整个公共区域有任何视觉上的实质阻隔。玄关也自然的拉长到底与餐厅合并再进一步和厨房串连。

客厅电视墙的下边刻意做了开口，和主卧的书房结合起来作为一个视觉上能够呼吸的透气口。主卧空间也把墙面全都打开了，从下方连接了浴室里的冲淋房，更衣室和书房也因为没有实质的隔墙看起来更为通透，地下室同样呼应了一楼的调性。

PROJECT
Natural Organic Space
DESIGNER
Hank M. Chao / MoHen Design International
COMPLETION
2008
PHOTOGRAPHER
MoHen Design International/Maoder Chou

项目
自然有机空间
设计师
牧桓建筑+灯光设计
完工时间
2008
摄影师
周宇贤

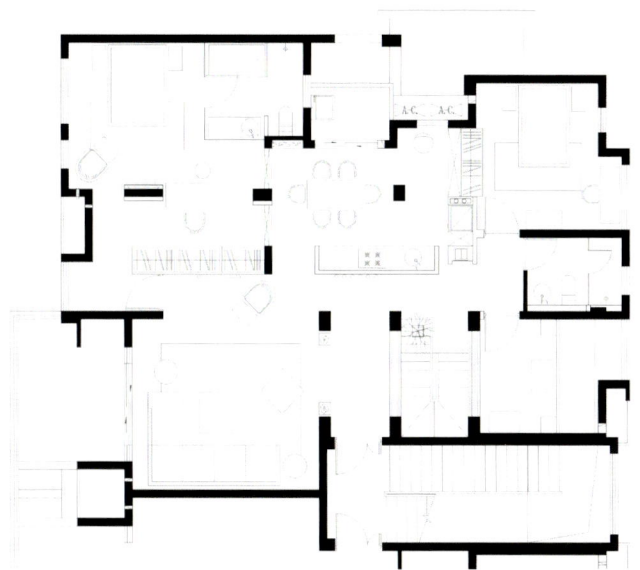

ORCHARD TURN SHOWFLAT
Singapore, Singapore

乌节弯公寓
新加坡 新加坡城

This show flat unveils a dream world of luxury residential comfort. Wood, stone and water features flourish in the U-shaped apartment, grounding it with an earthy, homey glow that well complements the sweeping panorama of Singapore.

On the left, a spacious open-plan kitchen stands close to the dinning room, allowing guests to interact and mingle as canapés and drinks are prepared on the walnut counter top. In the master bedroom, transparent glass is set on the surface of a silver travertine wall, subtly transpiring the mesmerizing grains of the marble beneath.

这个时尚的设计正揭示梦想中的现代型家居典范。设计师以木材、石材及水流作主导，配合狮城的繁盛全景，把自然优雅风格完完全全引入生活之中。

开放式厨房跟饭厅紧密连接，胡桃木柜台上可放置开胃菜和饮品，让来宾轻松自在地享用。主人套房内，透过清澈的玻璃看进银窿石的纹理，若隐若现，构成层次丰富的主题墙。

PROJECT
Orchard Turn Showflat
DESIGNER
Steve Leung Designers LTD
COMPLETION
2006

项目
乌节弯公寓
设计师
Steve Leung Designers LTD
完工时间
2006

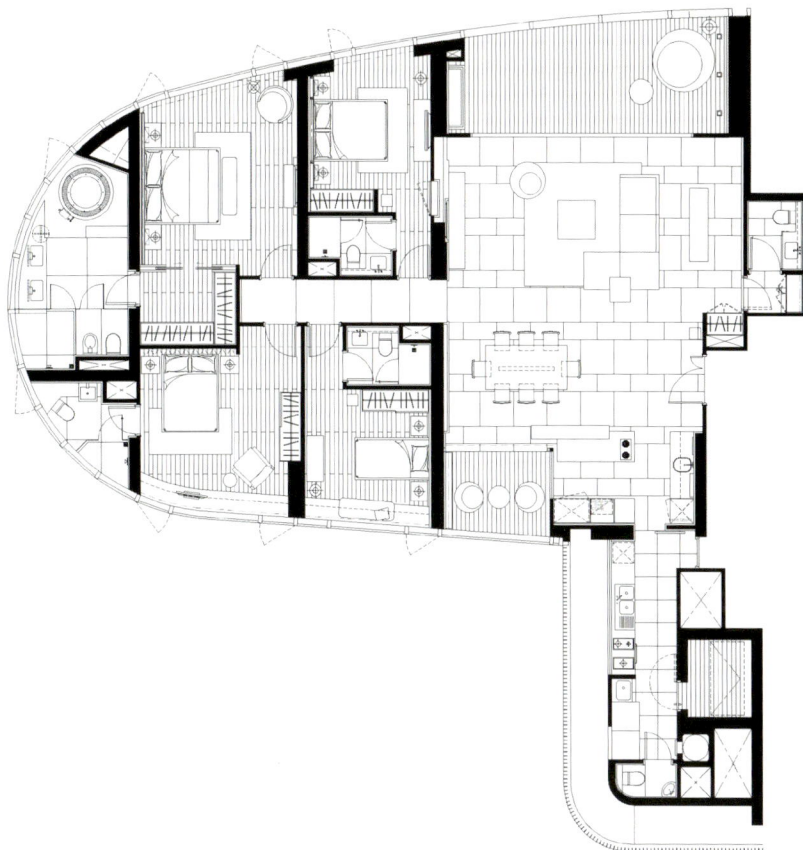

PEAK HOUSE
Hongkong, China

现代几何建筑美
中国 香港

As the recipient of the Andrew Martin Award, this house delights in the play of light that floods its interiors. As an eclectic blend of classical elements within a contemporary evironment, rooms flow into one another to create a series of surprising vistas. Relaxation is encouraged with stainless steel glimmering against a white palette that soothes and calms the senses.

A floating triangular staircase connects the three floors, with its profile outlined by treads in a dark wood stain. Up above, an arc of skylight brings in an unexpected ray of sunshine, enhancing the home's sculptural qualities.

现代几何建筑美是Cream 芸芸家居设计中之得意杰作，曾荣获英国著名奥斯卡室内设计大奖。看似一室清雅简洁的白及自然色系，却玩味几何形体的建筑线条及光线的运用，带出不平凡的现代家居设计。

当中最能表现室内建筑线条几何图形美的是贯穿这三层高洋房的一道扶手楼梯，中间玻璃天井部分呈等边三角形，与梯级的锯齿式三角形木块饰边互相呼应。同时，一层的落地大玻璃窗引入天然光线，与清雅的室内设计色彩共同塑造出居所的自然和谐美。

PROJECT
Peak House
DESIGNER
CREAM Design Company
COMPLETION
2003

项目
现代几何建筑美
设计师
CREAM Design Company
完工时间
2003

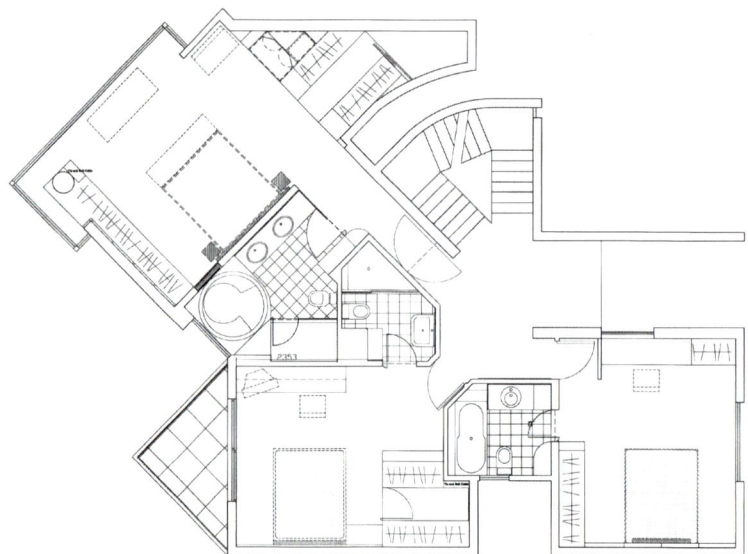

PENTHOUSE SCHRADER
Frankfurt, Germany

Schrader豪华公寓
德国 法兰克福

Hollin + Radoske designed this apartment all in white with a few contrasts in graphite grey and grey oak. Nearly everything is hidden behind built-in doors.

Four dressing rooms provide specially-designed space for the clothes collection of the owner, a lady with a particular interest in haute couture. The layout of the apartment is open and fluent; the whole glass façade is visible in the huge living and dining area. The kitchen comes as a hanging frame integrated in the white core of the apartment. The bedroom follows the same concept, highlighted with a few crisp colours, such as purple and violet.

The en suite bathroom fits perfectly with the calm and distinctive concept as it shows a grey-brown stone exactly reflecting the grey tinted oak from the sideboards and the kitchen. Backlit walls and indirectly-illuminated niches create a spacious and extravagant look.

这幢公寓以白色作为设计基调，另有少许黑灰色和灰色木纹点缀其中。貌似这背后藏匿着许多东西，而公寓的门把它们全部挡在了人们的视线之外。

四间更衣室经过特殊设计，足以容纳酷爱高档时装的女主人那为数众多的衣服。公寓的布局开放流畅，起居室和餐厅采用了一整面通透的玻璃幕墙。设计师把餐台做了雕刻艺术加工，在厨房中央的烹调区也使用了同样的手法。厨房如同一幅引人注目的艺术画作，让这幢白色的公寓变得更加丰满。卧室选用了清新的色彩，如深红色和紫色等，同样起到了画龙点睛的作用。

浴室套间的设计完全贴合平静独特的理念，浴室内的灰褐色石材恰恰与餐具柜和厨房的灰色橡木纹相呼应。背光式墙壁和起到间接照明作用的壁龛共同营造出宽广无限的效果。

PROJECT
Penthouse Schrader
DESIGNER
Hollin + Radoske Architects
COMPLETION
2005
PHOTOGRAPHER
Katharina Gossow

项目
Schrader豪华公寓
设计师
Hollin + Radoske Architects
完工时间
2005
摄影师
Katharina Gossow

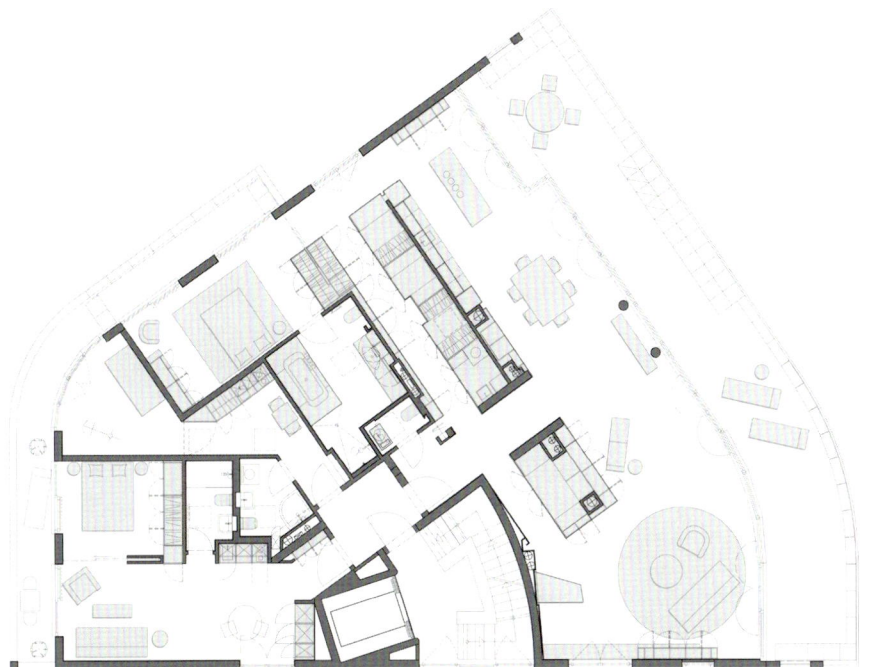

PRIVATE RESIDENCE TIP TOP LOFTS
Toroto, Canada

豪华公寓
加拿大 多伦多

The designers retained the back-to-back open kitchen and the narrow bathroom, to which they added a custom vanity with a white granite vessel basin, and marble cladding. Above these areas, II BY IV took advantage of the extraordinary ceiling height by creating a small mezzanine space, reachable by a fixed ladder, which provides additional storage, a reading nook and a guest sleeping area.

The scheme was grounded by the strong graphical effect of the oiled walnut strip flooring. In new partitions on either side of the bathroom, tall pocket doors with elegant hardware provide access to the living room and the bedroom, which is separated from the entrance hall by a wall of milky glass. That wall is centred on a full height walnut storage unit that provides double hanging, shoe organisation and storage above on the hallway side, and an integral dresser and double hanging on the bedroom side. Tidy, efficient and beautiful, the millwork unit, with its concealed hardware, reflects the owner's preference for minimalism. There is much concealed hardware through the

space. Against a sleek ten-foot tall, halo-lit, walnut headboard, the bed platform contains ample storage and is flanked on one side by a cantilevered white marble table and on the other, a white marble plinth upon which stand tall spindle sculptures by Dennis Lin.

设计师保留了原有的开放式厨房和浴室，并增设了白色花岗岩面盆和大理石台面。他们充分利用了空间的高举架，搭建出一个小型的阁楼，并设有通往阁楼的楼梯，阁楼上有储藏室、书房和客卧。

胡桃木地板为空间增添了丰富的图案效果。浴室两侧高大的房门分别通往客厅和卧室。一道乳白色的玻璃墙将卧室与门厅隔开，墙的中间是落地式的胡桃木储物柜。储物柜采用特殊设计，朝向门厅的一侧是鞋柜和衣柜，而卧室的一面则是梳妆台和双层衣柜。储物柜的设计简洁、实用、美观，与住宅整体风格协调一致，迎合了医生房主的偏好。卧室内的胡桃木床头高达3米，床内是储藏空间，床的一侧是白色大理石边桌，另一侧的白色大理石基座上伫立着丹尼斯·林设计的雕塑。

PROJECT
PRIVATE RESIDENCE TIP TOP LOFTS
DESIGNER
II BY IV Design Associates Inc.
COMPLETION
2006
PHOTOGRAPHER
David Whittaker

项目
豪华公寓
设计师
II BY IV Design Associates Inc.
完工时间
2006
摄影师
David Whittaker

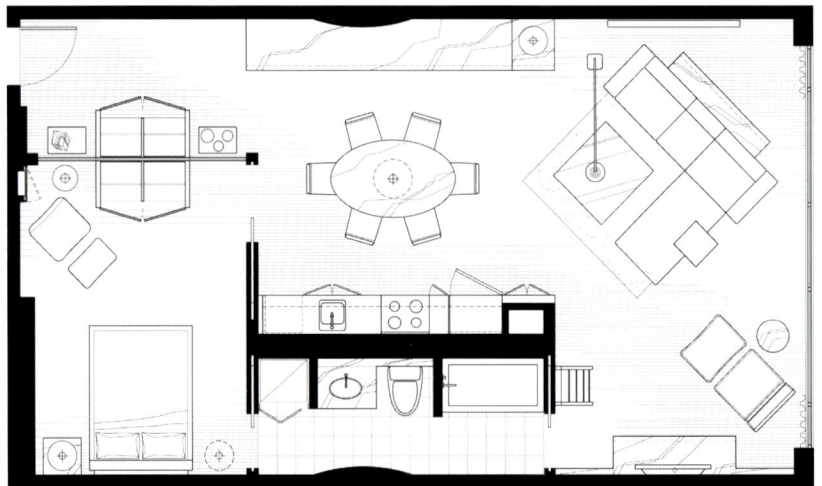

QUANT
Stuttgart, Germany

船桨之家
德国 斯图加特

The concept is about creating a living environment that goes far beyond the average and approaches what living is really about – getting the most out of life. Similar to a loft space, the apartment is immediately tangible as a generous, continuous space. It can be apprehended in its entirety from certain vantage points. All functions are accommodated in freely-defined areas, which can be closed off by means of sliding doors and heavy curtains, if so desired. In this way, a whole range of new and beautifully-framed interior and exterior vistas become apparent.

The private part of the apartment is designed around the sleeping and working areas. The bedroom and the study form a common zone, defined by a circle of oak, encasing wall, floor and ceiling.

The flowing room concept is supported by an ingenious composition of materials and colours – the elegant, olive green, epoxy resin-coated floor, oak wood surfaces, white varnished finishes, sensual fabrics and coloured walls evoke

an exciting dialogue with one another and, of course, with the apartment's occupants.

整个设计打造了一个与众不同的居住空间，最大限度地接近居住的真谛——取得最大的人生意义为原则。这个房间宽敞、明亮，很像一个LOFT。从某一个合适的角度观察，房间浑然成一体。不同的功能区都以滑动门或厚窗帘为隔断。通过这种方式，使悦目的内景和外景都清晰可见。

这个房间的私人空间主要围绕休息空间和工作空间进行设计。当客人参观房间的时候，你会陶醉于他们艳羡的目光中。卧房和书房之间被橡木墙、地板、天花板围绕。

材料和颜色的独特组合完美诠释了"流动的房间"这一概念：考究的橄榄绿色环氧复合材料地板、橡木、白色罩面漆、富有质感的织物、与多彩的墙面正进行一场生动的对话，连房间的主人都参与其中。

PROJECT
QUANT
DESIGNER
ippolito fleitz group
COMPLETION
2005
PHOTOGRAPHER
ZOOEY BRAUN

项目
船桨之家
设计师
ippolito fleitz group
完工时间
2005
摄影师
ZOOEY BRAUN

RED
Hongkong, China

红
中国 香港

The flaming red energises the main room, and the white colour with glass and decorative mirrored walls is extremely impressive in the whole interior.

The emphasis of the living room, with its mirrored glass wall decorated with rounded shapes of red and its contrasting effect with white, makes a stunning effect. Touches of fiery colours are carried out throughout the corridors as well as the household appliances which show a flamboyant charm.

The design blends the mood of the 1970s with contemporary styles, creating a dramatic yet comfortable ambiance.

这个项目以"红"为主题，整个室内设计强调一室的白、落地玻璃和装饰性镜墙，在冷峻中吐露火艳的红。

一步入客厅，点缀着一圈圈圆镜片的红色光由玻璃装饰墙映入眼帘，与一室的白形成强烈的对比，充满震慑的效果。而红色的点缀亦贯彻于走廊上的方形图案镜墙和家居饰品之上，发挥火红的妖艳魅力。

此外，在充满现代绮丽感的室内设计下，透过落地玻璃和装饰性镜墙上的几何图案，注入一点20世纪70年代的风情，产生一种新旧交融的设计效果。

PROJECT
RED
DESIGNER
CREAM Design Company
COMPLETION
2006

项目
红
设计师
CREAM Design Company
完工时间
2006

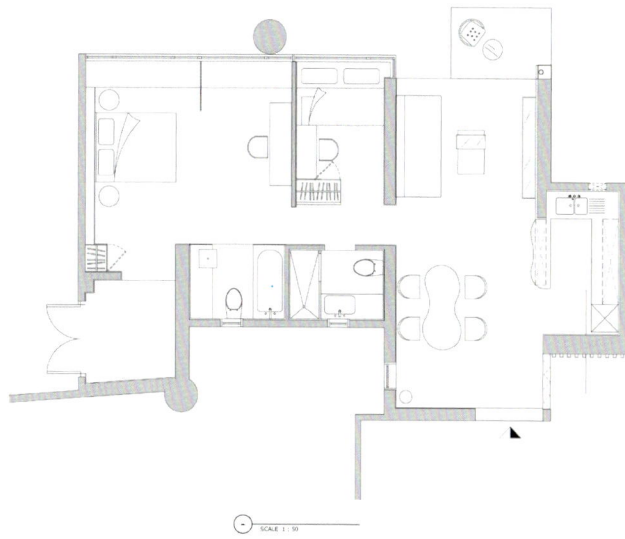

SCALE 1 : 50

RESIDENCE TSAI'S
Taipei, China

敦南国美蔡宅
中国 台北

The project is a combination of two units on Dunhua South Road in Taipei. Based on the request of the owner, the entire space was redivided into two sections: the master bedroom section for everyday life and the guestroom section for the parents who visit the owner occasionally. It is further bent and folded into a U-shaped space which becomes the bathroom.

A huge piece of glass between the living room and the bedroom maintains the continuation of the two spaces. The curtain is the main element to separate these rooms. When the curtain is down, the bedroom becomes a private space, as if it is a space inside space. Two blocks of different attributes on the left and right are integrated in the 3D ceiling whose curve changes gradually from the guestroom to the children's room, from gentle to twisted, creating the feeling of a cave or a hollow space. Pipelines of the air-conditioner and lighting are embedded in the twisted 3D ceiling to create more space. The multidimensional stone-cut and diamond-cut counter top of the kitchen set creates the sense of an artificial stone island. Furniture decorates the space and balances the pure style

of the original space, and also brings out the designers' taste of life.

该项目是敦化南路上一间被打通而合并的住宅空间，根据业主使用需求重新规划两大区域，分成日常生活起居使用的主卧房区和给父母及亲属使用的客房区。主卧房区除了一个豪华的主卧房，更衣室与浴室之外还预留一间以后可作为儿童房的书房和小吧台。

客厅与卧房中间使用玻璃隔断，平时可利用窗帘来控制空间的对应关系，拉上窗帘大卧房就形成独立区域，好像是空间里的空间。左右两个不同属性的区域，则利用三度天花板统合，由客房区到孩童房间，天花板的曲线渐变，从平缓到扭曲，所有波浪都是倾斜的，有点像挖出的洞穴空间。用平面制造3D洞穴感，另将空调管线与灯光等功能藏入扭出的3D天花板里以争取其他地方的空间高度。

客厅及餐厅则连续成为一个大空间，电视柜与玄关处的鞋柜合并成一柜，两面作为隔屏，沙发独立；厨房则做成中岛空间，使用多角型石头、钻石切割面造型的人造石岛台，家具点缀了空间，均衡原空间纯粹的造型，也增添了生活的品位。

PROJECT
Residence Tsai's
DESIGNER
Shichieh Lu
COMPLETION
2007
PHOTOGRAPHER
Marc Gerritsen

项目
敦南国美蔡宅
设计师
陆希杰
完工时间
2007
摄影师
马克·盖瑞森

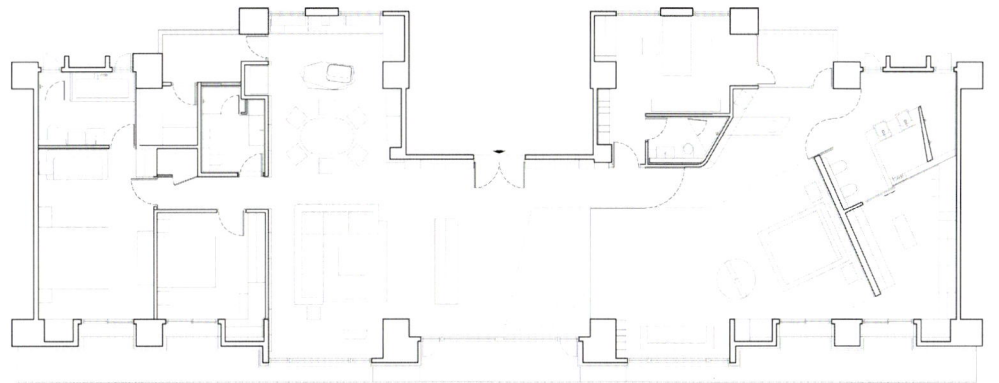

SOUTH BAY
Hongkong, China

南湾
中国 香港

As the recipient of the Andrew Martin Award, this three-storey home contains a wealth of spaces that make it an ideal abode for entertaining. It possesses an elegance that is enhanced by a staggered display case with recessed lighting in the living area.

A central light well acts as a unifying focal point, with the staircase wrapping around this gloving niche. While the public areas are more neutral and feature furnishing such as a quartet of white marble coffee tables, the private areas – the bedrooms, for example –contain bold matches of colours.

Meanwhile, a lap pool and an outdoor barbecue area allow for outdoor entertainment. The dominant feeling is luxury, with a dash of eclecticism to keep things interesting.

这座位于南湾的多空间三层独立洋房是一处很不错的住所，曾荣获英国著名奥斯卡室内设计大奖。其错落有致的陈设设计和隐秘的照明安排使整个空间气质幽雅。

楼梯环绕着中央灯饰，使其很自然地成为室内的焦点。公共空间含蓄典雅的设计，由四个独立倒转梯形云石组合而成的咖啡桌，到卧室和其他的私人空间均流露室内设计细节的非凡之处。户外的泳池和露天烤肉平台为室外增加了娱乐空间。

室内设计整体感觉奢华亮丽，设计师用折中的理念为这里增添了有趣的意境。

PROJECT
South Bay
DESIGNER
CREAM Design Company
COMPLETION
2003

项目
南湾
设计师
CREAM Design Company
完工时间
2003

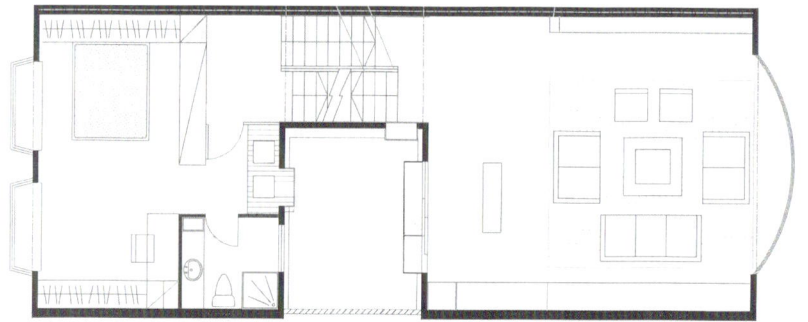

SPACE OF FEMININITY
Shanghai, China

女性居家环境
中国 上海

This small space was designed for a free, single woman. The designer was inspired to use curvy lines to present the space, believing that simple arcs and curves make a space look more lively than not. Even if you don't plan colours and materials, a space can look good. However, adding a large arc or a line can ruin it, if not used correctly, while using curves for details appropriately can sometimes add a little excitement to an area. Therefore in this project, curves were added to the door frames, the main wall of the living room, the bed heads and the furniture, giving the whole space a little more characters.

Having a colourful wall in the hallway and putting some eggshell-like chairs in front of it give a clear indication of the owner's age and sex. Having no walls in between the living room and the dinning room gives a feeling of a lounge bar to the guests. Besides, there is a lively atmosphere, created through a few piled circles which separate private and public spaces with a bed in air, presenting a modern home of a beauty.

这个小尺度的空间是为单身自由的女性设计的。设计师用曲线来解释这个空间，轻巧活泼的弧形和曲线能让空间显得更加轻盈，配色和用料大胆一点也变得合情合理。但如果在平面上使用大弧形的这种较大的装饰性语言就显得太过了，反倒在小巧一点的地方，东一点西一点的让空间有某种程度上的小对比更能让人有些意外的惊喜。因此，在门框、客厅主墙面、床头背板、家具……这些地方的小弧形和小线条让空间能更可爱一些。

在入口门厅，放了一个色彩艳丽红红粉粉的装饰墙面，前面再摆些白色带黄像鸡蛋壳般的小座椅，清楚传递了住在这里的使用者的性别和年龄阶层。客厅与餐厅没有隔间墙，让两边空间能更紧密，视觉上也有连接，客人来时在这样的开放空间里倒是有点Lounge Bar的气氛，但是属于比较活泼热闹的那种生动感。用几个圆圈交叠的墙界定私密与公共空间，卧室的床悬挑飘在半空中 —— 一个现代式的美人窝。

PROJECT
Space of Femininity
DESIGNER
Mohen Design International/Mohen Chao
COMPLETION
2006
PHOTOGRAPHER
MoHen Design International/Maoder Chou

项目
女性居家环境
设计师
MoHen Design International/赵牧桓
完工时间
2006
摄影师
MoHen Design International/周宇贤

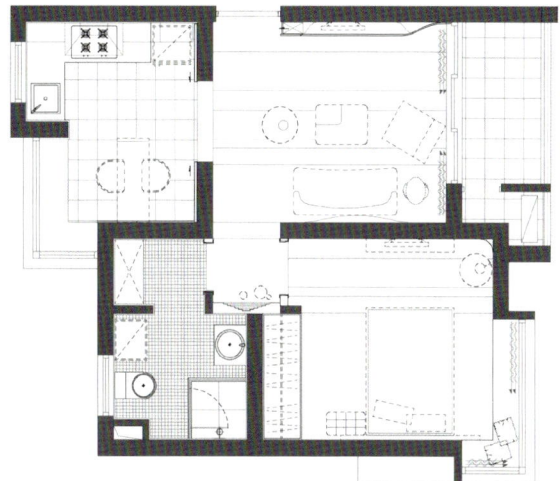

SPRING HILL HOUSE
Queensland, Australia

春山之屋
澳大利亚 昆士兰

Passive climatic design was integral for ensuring adequate natural light, shading and ventilation throughout the house. The roof form of the bedroom wing incorporates the dual ideas of a lower-pitched verandah roof and window hoods. This results in a pitched roof with a distinctive off-centre ridge that acts as a sun and weather protection on the north facing the wall and the windows. Window seats have been incorporated to provide sun shading, privacy and to frame the city views. The bedroom wing has been positioned to encourage natural light and ventilation into the main living areas of the house.

The covered outdoor living spaces provide a connection with the landscape, and take advantage of the subtropical climate and the city views. The integration of vegetation for privacy and shading, the terraced lawns and the swimming pool enhance the livability and quality of the landscape areas. The outdoor living areas are positioned to capture cooling breezes for cross ventilation and natural lighting.

设计顺应了气候的要求，使房子里有充足的自然光，其避光处和通风效果也恰到好处。卧室的侧翼有两种用途，楼下后期搭建的阳台以它做天花板，并且卧室的小窗也有了窗檐。这样，阳台的天花板明显的偏离了中心屋脊，探出的部分还保护了北面的墙壁和窗户免受阳光和恶劣天气的侵袭。窗边的座位设在阴凉处，兼顾私密性的同时又能观赏城市景观。卧室侧翼将更多的自然光照和流通空气引入主要的起居空间，同时仍兼顾着视野和隐私。

非露天的室外起居空间充分利用了当地的亚热带气候和城市景观，是与户外美景的绝佳连接。一体化的植被既照顾了隐私，又能供人乘凉，阶梯式草坪和游泳池提高了房子的可居住性和景观质量。在室外起居空间，能同时享受凉爽的微风和自然光照。

PROJECT
Spring Hill House
DESIGNER
8i Architecture Pty Ltd
COMPLETION
2005
PHOTOGRAPHER
Taylormade Digital Images

项目
春山之屋
设计师
8i Architecture Pty Ltd
完工时间
2005
摄影师
Taylormade Digital Images

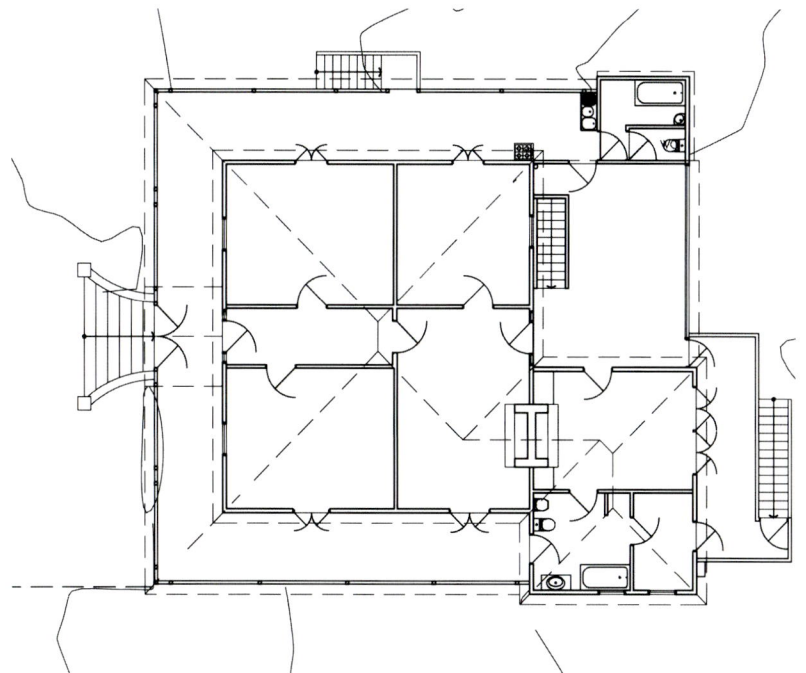

STONEHEDGE RESIDENCE
Austin, USA

斯通家园
美国 奥斯丁

PROJECT
Stonehedge Residence
DESIGNER
Miró Rivera Architects
COMPLETION
2004
PHOTOGRAPHER
Paul Finkel, Piston Design

项目
斯通家园
设计师
Mir ó Rivera Architects
完工时间
2004
摄影师
Paul Finkel, Piston Design

The entrance was reconfigured to create a clear procession, which didn't exist before. The dining room was co-opted to create a foyer suitable to the house with a large pivot door clad in copper tubing welcoming visitors. A natural stone path was selected to integrate the garden with the new entrance to the house.

The great room was expanded in volume to establish a clear hierarchy of spaces in the house. This was achieved by capturing the space of the existing screened porch and by eliminating the floor above to create a double-height space. The great room is anchored on one side by a 19-foot-copper-clad fireplace. On the opposite side, the room flows to the outside terrace through a glazed wall with large sliding doors.

This large room serves as a hub for the entire house, with all the functions of the house revolving around it – to the north the entrance, to the south the terrace, to the west the bedrooms, and to the east the kitchen, the garage and the home offices. The terrace and the trellis were created to unify the house and the garden.

这个完成的设计项目通过三个主要措施，达到了双赢：

设计师把入口改造得十分明朗，与之前大不相同。餐厅处增设了一个与房子规模相匹配的门厅，一扇巨大的包铜转门在那里迎接来宾。选用天然石铺设甬道，使节水型园艺花园与房子的新入口完美的结合。

大房间变得更加开阔，提高了房子的空间层次感。设计师利用了原先屏风式门廊的空间，并拆除了二楼，才创造出双层高度的空间。大房间一侧是19英尺高的包铜壁炉。对面是玻璃幕墙和一扇大拉门，直通外面的阳台。

大房间是整幢房子的枢纽，房子的所有功能性区域都围绕在它的周围。它的北面是入口，南面是阳台，西面是卧室，东面是厨房、车库和书房。阳台和格架拱道把房子与花园和谐的统一起来。

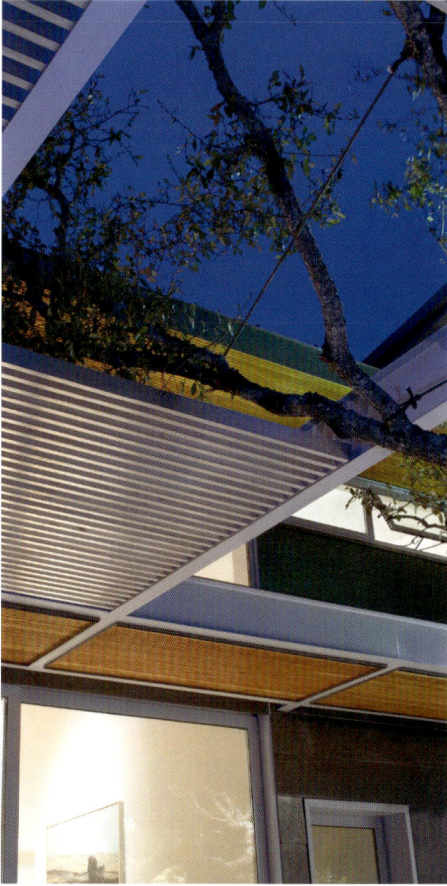

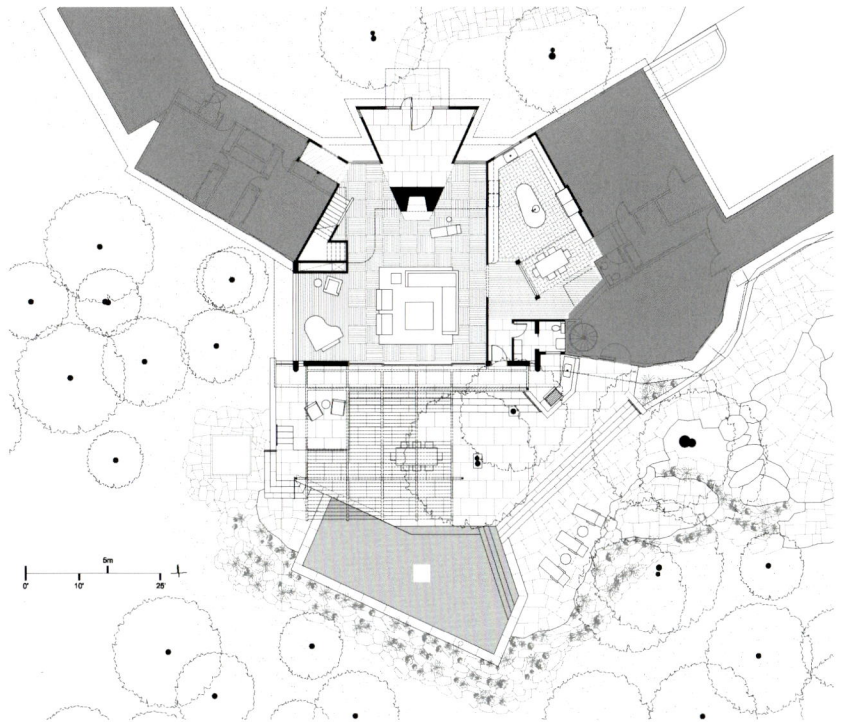

TOWN LANE RESIDENCE
NewYork, USA

乡间小巷住宅
美国 纽约

The house is designed to accommodate an active family with four children. On the ground floor is one large room containing a living room, a dining room and a kitchen similar to what one can see in a city loft. A large screen porch is tucked into one corner of the plan. In summer the porch is a hub of activity; in winter, it doesn't obstruct light or views.

The three-storey stairwell connects all floors. The basement playroom, with access to outdoors and the pool, does not feel or function as a typical cellar recreation room. In fact, it is where everyone comes together to watch movies and play table tennis. Natural light pours in through a wall of sliding glass doors.

The bedrooms are on the first floor, all with views of the ocean. An outdoor deck connects the master bedroom to a trellised hot tub area with an outdoor shower. Sunlight streaming through the lattice casts dynamic shadows and makes this "house" a favourite place for family fun.

房子的设计可以容纳一个拥有四个孩子的活力十足的家庭。一楼是一个大房间，包括起居室、餐厅和厨房，和一般的城市Loft差不多。巨大的屏风式门廊被安置在角落里。到了夏天，门廊就是主要的活动区。在冬天，它又不会阻挡光线和视野。

三层楼梯井贯通所有楼层。地下游戏室直通户外和游泳池，感觉和一般的地下康乐室很不一样。事实上，一家人可以聚在这里看电影和玩乒乓球。自然光能从滑动玻璃门中倾洒进来。

卧室在二楼，全都可以看到大海。一个露台连接着主卧室和搭建而成的浴室，可以享受室外淋浴。阳光透过栅格窗照射进来，形成流动的光影，把这座房子变成了充满家庭乐趣的爱宅。

PROJECT
Town Lane Residence
DESIGNER
DONALD BILINKOFF ARCHITECTS
COMPLETION
2005
PHOTOGRAPHER
DONALD BILINKOFF ARCHITECTS

项目
乡间小巷住宅
设计师
DONALD BILINKOFF ARCHITECTS
完工时间
2005
摄影师
DONALD BILINKOFF ARCHITECTS

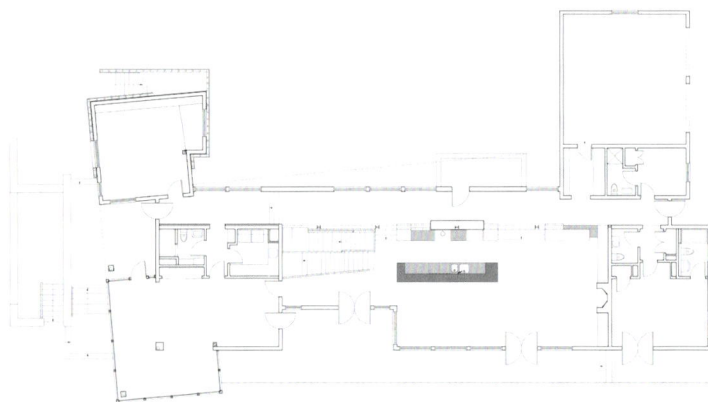

UPDOWN COURT SHOWFLAT
Taiwan, China

爱敦阁样板房
中国 台湾

The designer's mission is to make the show flat attractive enough. Elegant silver and grey are chosen as the primary colour scheme in the design.

Glass and mirror are shiny and sparkling and a large amount of tinted glass and mirror is used here to achieve the effect and to enhance the visual impact.

Geometric forms and symmetric lines are classic, elegant and timeless. The most successful use of them is in the Art Deco period artworks. The whole design of the project is thus inspired.

Wall panels with vertical asymmetrical groove lines are installed on the opposite wall and are extended to the TV wall. The groove lines would gradually be transformed to abstract floral patterns to decorate the wall.

设计师的使命是让样板房充满致命的吸引力。设计中，他们选用优雅的银色和灰色作为主色调。

营造豪华氛围的关键是利用有光泽和闪闪发光的元素。玻璃和镜面正符合这两点要求，因此我们使用了大量的有色玻璃和镜面，以达到预期效果，提高视觉冲击力。

几何图形和整齐的线条是如此经典、高雅和永恒。装饰艺术运动时期的艺术品是对它们最成功的运用。我们设计该项目的全部灵感正来源于此。

在墙面上开槽，挖出垂直但不对称的线条，让它们一直延伸到电视墙上。这些线条逐渐演变成抽象花朵的形状，起到墙面装饰的作用。

PROJECT
Updown Court Showflat
DESIGNER
Ching Ping CHANG, Cherry TANG, Louis LAW, Chun Ern YEH, Yu Cheng Wang, Yu You Liao
COMPLETION
2007
PHOTOGRAPHER
Shou Shan LAI

项目
爱敦阁样板房
设计师
Ching Ping CHANG, Cherry TANG, Louis LAW, Chun Ern YEH, Yu Cheng Wang, Yu You Liao
完工时间
2007
摄影师
Shou Shan LAI

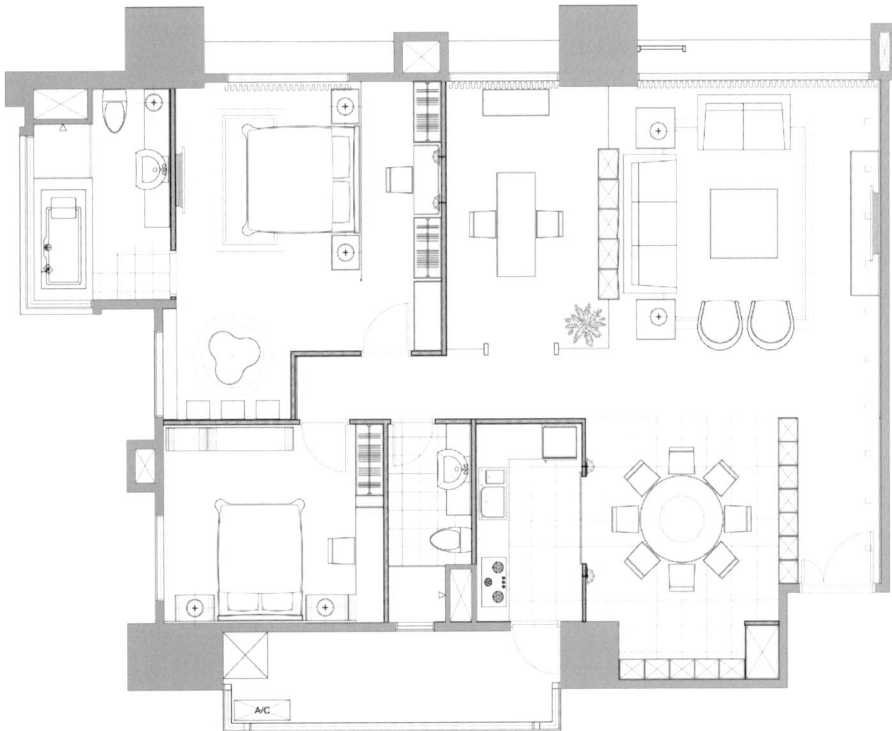

WINDSOR LOFT
Victoria, Australia

温莎阁楼
澳大利亚 维多利亚

The design is a reflection on architectural strategies – spatial intervention/differentiation in a fluid continuity; and visual/subliminal permeation between private and public domains in contemporary domestic spaces in urban fabrics. A cruciform column stands in the middle of the ground floor living area. Traversing floor tiles and recessed ceiling lighting tracks radiate from the four edges, virtually dividing the open plan into entertaining, dining, sitting and circulation areas.

With its wall-lined, high-gloss, white joinery and the north orientation to the internal courtyard, the kitchen blurs into an inconspicuous whitewash in daylight. At night it becomes an annex to the house.

As a separate but un-partitioned wing of the main bedroom, the en suite room is designed as a small conservatory. In daylight the mirror wall reflects its entire surroundings – the sky, the bedroom, the courtyard and the living area beyond – so much that the physical structures are seemingly dispersed, and the vanity basin, the bath and the shower rose become part of the reflected images.

PROJECT
Windsor Loft
DESIGNER
Architects EAT
COMPLETION
2006
PHOTOGRAPHER
John Gollings, Shania Shegedyn, Jason Reekie

项目
温莎阁楼
设计师
Architects EAT
完工时间
2006
摄影师
John Gollings, Shania Shegedyn, Jason Reekie

设计立足于建筑视角——在现代城市建筑的室内设计中使用连续的空间介入与分割，并在私人空间和公共空间中进行视觉与潜意识的渗透。一楼大厅中央竖立着一根十字形圆柱。它下端横跨地砖，上端伸入天花板的灯光中，并投下四条暗影，明确地划分出娱乐、餐饮、起居和公共区域。

顺着墙边高大的亮白色木材装饰向北走，进入室内庭院，日光下的厨房覆盖着不起眼的白色涂料，竟险些被忽视。到了夜晚，它立刻变成一个夺目的亮点。厨房里的红色半透明操作台亮了起来，调色灯光为酒吧增添了一抹绯红的色彩，昏暗中现出弯弯曲曲的光。

客用浴室和客厅之间由一层红色玻璃隔开。尽管玻璃并不透明，但人们对玻璃的印象是不能完全保证隐私，这时主观意识让私人空间与公共空间的界限也变得暧昧了。

浴室是主卧室一个独立的区域，却没有被彻底分离出去，它设计得更像一个小暖房。白天，镶有镜子的墙面把它周围的一切事物都映射出来，包括天空、卧室、庭院、起居室等，包罗万象看似杂乱无章，漱洗台、浴缸和淋浴器也成为映像中的一部分。

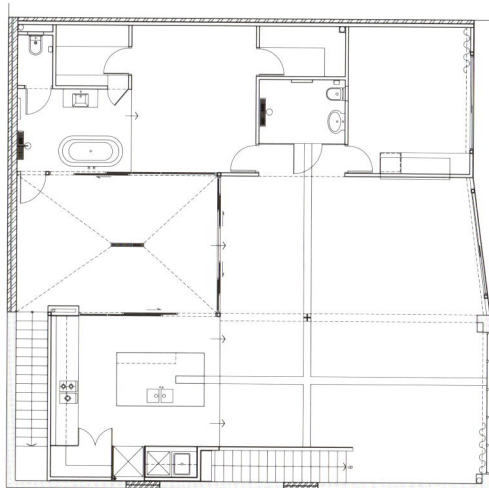

Index 索引

Tien fun Interior Planning Co., Ltd.

http://tienfun.com.tw/

T: (886) 4-2201 8908

Cream design

58 kings Road
Reading
RG1 3AA

T: 0118 958 6946
www.cream-design.co.uk

8i Architecture Pty Ltd

21 Hansen Street

T: (07) 3342 4806
Moorooka Qld 4105
F: (07) 3342 4807
ABN 75 279 050 454
mail@8i.net.au
www.8i.net.au

DONALD BILINKOFF ARCHITECTS

310 Riverside Drive
Room 202-1
New York, NY 10025

T: (212) 678-7755
F: (212) 678-7743

ippolito fleitz group identity architects

Bismarckstraße 67b D-70197 Stuttgart

T: +49 (0)711 993392-335
F: +49 (0)711 993392-333
info@ifgroup.org
www.ifgroup.org

GRAFT

Gesellschaft von Architekten mbH
Heidestrasse 50   10557 Berlin   Deutschland

T: 030 / 2404 79-85 oder -86
F: 030 / 2404 79-87
berlin@graftlab.com
www.graftlab.com

II BY IV Design Associates Inc .

77 Mowat Avenue, Suite 109
Toronto, Ontario
M6K 3E3

T: 416 531 2224 x 225
F: 416 642 0102
www.iibyiv.com

Jsa Design Studio

Culiacan 123, 6th&7th floor
Hipodromo Condesa
06170 Mexico City

T: +52(55)10859900
www.jsadd.com

Mohen Design International

No.18,Alley 396, Wulumuqi S. Rd.
Shanghai, China 200031

T: +86-21-64370910/64374175/64374462
F: +86-21-64317125
mohen@mohen-design.com
www.mohen-design.com

SHH architects + interiors + design consultants

1 Vencourt Place
Ravenscourt Park
Hammersmith, London
W6 9NU

T: 020 8600 4171
F: 020 8600 4181
http://www.shh.co.uk

Ptang Studio Ltd.

Rm 603-604, Harry Industrial Building,
49-51 Au Pui Wan Street, Sha Tin,
New Territories, Hong kong

T: (852)26691577
F: (852)26693577
www.ptangstudio.com

ESPERTA architecture-interior

JI Artha Gading Blok A6B No 16
Kelapa Gading – Jakarta
Indonesia
14240

T: +62-21-45873976
F: +62-21-45873975
E: esperta@cbn.net.id

Miró Rivera Architects

505 Powell Street
Austin, Texas 78703

T: 512 477 7016
F: 512 476 7672
info@mirorivera.com
www.mirorivera.com

8i ARCHiTECTURE

Moorooka, Qld, 4105

T: +61-7-3342-4806
F: +61-7-3342-4807
www.8i.net.au
mail@8i.net.au

Hollin+Radoske Architects

Goldsteinstrape 61a
60528 Frankfurt/Main
Germany

T: +49(0)69-49449890
F: +49(0)69-494498929
info@hollinRadoske.de
www.hollinradoske.de

PTang Studio Ltd

rm 603-604, harry industrial building,
no. 49-51 au pui wan street, shatin
Hongkong

T: 2669 1577
F: 2669 3577
office@bipt.com

Kinari Design

2 Palace Gardens Terrace, London, W8 4RP

T: 020-7221 0082
doug@kinaridesign.com
www.kinaridesign.com

GFAB Architects

Jl. Kertha Usadha III / 24-28, Denpasar Bali
80224 Indonesia

T: +62 361 723195 / 720296
F: +62 361 726562
www.gfabarchitects.com

Hecker Phelan & Guthrie P/L

1 Balmain Street, Richmond Victoria 3121

T: 03 9421 1644
F: 03 9421 1677
www.hpg.net.au

Donald Billinkoff Architects

310 Riverside Drive, Suite 202-1,
New York, NY 10025

T: 212 678 7755
F: 212 678 7743
www.billinkoff.com

GREG NATALE DESIGN

Studio 6 level 3 35Buckingham Street Surry
Hills NSW 2010

T: 02 8399 2103
F: 02 8399 3104
www.gregnatale.com

Steve Leung Designers LTD

9/F Block C Seaview Estate 8 Watson Road
North Point Hongkong

T: 852 2527 1600
F: 852 2527 2071
www. steveleung.com

CJ STUDIO

Floor 6 No 54, Lane 260 Kwang Fu South
Road, Taipei, Taiwan

T: (02)2773-8366
F: (02)27738365
cj@shi-chieh-lu.com
www.shi-chieh-lu.com

Cha & Innerhofer

Architecture+Design
611 Broadway
New York, NY, 10012

T: (212) 477-6957

图书在版编目（ＣＩＰ）数据

精品屋／赵婷婷编. — 沈阳：辽宁科学技术出版社，
2010.03

ISBN 978-7-5381-6037-6

I. 大… II. 赵… III. 住宅－室内设计－作品集－世界
IV.TU241

中国版本图书馆CIP数据核字（2009）第158744号

出版发行：辽宁科学技术出版社
　　　　　（地址：沈阳市和平区十一纬路29号　邮编：110003）
印　刷　者：利丰雅高印刷（深圳）有限公司
经　销　者：各地新华书店
幅面尺寸：225mm×285mm
印　　　张：17
插　　　页：4
字　　　数：150千字
印　　　数：1~2500
出版时间：2010年03月第1版
印刷时间：2010年03月第1次印刷
责任编辑：陈慈良
封面设计：李　莹
版式设计：李　莹
责任校对：周　文

书　　　号：ISBN 978-7-5381-6037-6
定　　　价：228.00元

联系电话：024-23284360
邮购热线：024-23284502
E-mail: lnkjc@126.com
http://www.lnkj.com.cn
本书网址：www.lnkj.cn/uri.sh/6037